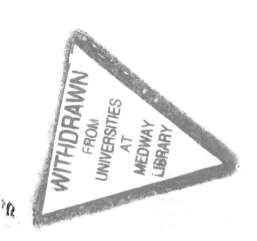

The Geological Society of London Handbook Series
published in association with
the Open University Press comprises

Barnes: *Basic Geological Mapping*
Fry: *The Field Description of Metamorphic Rocks*
McClay: *The Mapping of Geological Structures*
Milsom: *Field Geophysics*
Thorpe and Brown: *The Field Description of Igneous Rocks*
Tucker: *The Field Description of Sedimentary Rocks*

Professional Handbooks Series
Brassington: *Field Hydrogeology*
Clark: *The Field Guide to Water Wells and Boreholes*

Geological Society of London Handbook
HANDBOOK SERIES EDITOR — KEITH COX

# Field Geophysics

John Milsom

*Department of Geology*
*University College*
*University of London*

OPEN UNIVERSITY PRESS
Milton Keynes

and

HALSTED PRESS
John Wiley & Sons
New York — Toronto

Open University Press
12 Cofferidge Close
Stony Stratford
Milton Keynes MK11 1BY

First Published 1989

**British Library Cataloguing in Publication Data**
Milsom, John
  Field geophysics.
  1. Geophysics. Techniques
  I. Title   II. Series
  551′.028
ISBN 0-335-15207-4 ✓

Published in the U.S.A., Canada and Latin America by
Halsted Press, Division of John Wiley & Sons, Inc.,
New York

**Library of Congress Cataloging in Publication Data**
Milsom, John, 1939–
  Field geophysics.
  (Geological Society of London handbook)
  1. Prospecting—geophysical methods. I. Title.
  II. Series: Geological Society of London handbook
  series.
TN269.M53   1989      622′.159      88-19710
ISBN 0-470-21156-3 (Halsted) ✗

Printed in Great Britain by
Butler & Tanner Ltd, Frome and London

# Contents

# Preface

The purpose of this book is to help anyone involved in small-scale geophysical surveys. It is not a textbook in the traditional sense, in that it is designed for use in the field and concerns itself with practical matters — with theory taking second place. Where theory determines field practice, it is stated, not developed or justified. For example, no attempt is made to explain why four-electrode resistivity works where two-electrode surveys do not.

The book does not deal with marine, airborne or downhole geophysics, nor with deep seismic reflection work. In part this is dictated by the space available, but also by the fact that such surveys are usually carried out by quite large field crews, at least some of whom, it is to be hoped, are both experienced and willing to spread the benefit of that experience more widely.

Where appropriate, some attention is given to jargon. A field observer needs not only to know what to do but also the right words to use, and right in this context means the words which will be understood by others in the same line of business, if not by the compilers of standard dictionaries.

A word of apology is necessary. The field observer is sometimes referred to as 'he'. This is unfortunately realistic, as 'she' is still all too rare, but is not intended to indicate that 'she' is either unknown or unwelcome in the geophysical world. It is hoped that all geophysical field workers, whether male or female and whether geophysicists, geologists or unspecialized field hands, will find something useful in this book.

Finally, a word of thanks. Paul Hayston of BP Minerals and Tim Langdale-Smith of Terronics read early drafts of the text and made numerous invaluable suggestions. To them, to Janet Baker, who drew many of the sketches, and to the companies which provided data and illustrations, I am extremely grateful.

# 1

# *Introduction*

## 1.1 Fields

Although there are many different types of geophysical measurement, small-scale surveys all tend to be rather similar and involve similar, and sometimes confusing, jargon. For example, the word 'base' has three different common meanings, and 'field' and 'stacking' have two.

Measurements in geophysical surveys are made 'in the field'; unfortunately, many are also 'of fields'. Field theory is fundamental to gravity, magnetic and electromagnetic work, and even alpha-particle fluxes and seismic wavefronts can be described in terms of 'radiation fields'. Sometimes ambiguity is unimportant, and sometimes both meanings are appropriate (and intended), but there are occasions when it is necessary to make clear distinctions. In particular, the term 'field reading' is almost always used to identify a reading made *in* the field, i.e. not at a base.

Fields can be illustrated by lines of force which show their direction at any point. Closely-spaced lines indicate strong fields.

### 1.1.1 The inverse-square law

Inverse-square attenuation of signal occurs in most branches of applied geophysics. It is at its simplest in gravity work; the field due to a point mass is inversely proportional to the square of the distance and the constant of proportionality (the gravitational constant $G$) is invariant. Magnetic fields also obey an inverse-square law and although their strength is modified by the permeability of the medium, this is unimportant in geophysical work where measurements are made in either air or water. Complications due to the bipolarity of magnetic sources are much more significant.

Electric current flowing in a continuous homogeneous ground provides a physical illustration of the significance of the inverse-square law. All of the current injected into the ground at a point electrode must cross a closed surface around that point. If the surface is a hemisphere centred on the electrode, the current per unit area will be uniform and therefore inversely proportional to the surface area of the hemisphere, i.e. to the square of the radius. Flow patterns are, of course, drastically modified by inhomogeneities.

Seismic waves radiate from point sources as spherical wavefronts but are subject to absorption; the energy crossing a closed surface is less than the energy emitted by the source. Absorption also modifies radioactive flux intensities. Attenuation of electromagnetic waves generally follows an inverse-square law but can be modified by atmospheric absorption and by focussing.

depend on source shapes as well as on the inverse-square law. If the source 'point' in Fig. 1.1 represents an infinite line-source seen end-on, the area of the enclosing (cylindrical) surface is proportional to the radius and the field strength is therefore inversely proportional to distance and not to its square. This is roughly true of all infinitely-long sources of constant cross-section, termed 'two-dimensional' and providing reasonable approximations to bodies of large strike-extent.

### 1.1.2 *Two-dimensional sources*

Rates of decrease in field strengths

### 1.1.3 *One-dimensional sources*

The lines of force or radiation inten-

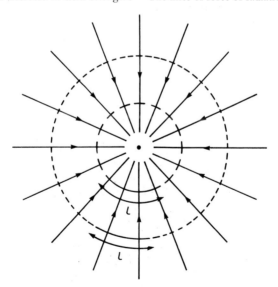

**Fig. 1.1** Lines of force from an infinite line-source viewed end-on. The line spacing increases in proportion to the distance from the source, so that the arc, length *l*, of the inner circle is cut by four lines but, on the outer circle of twice the radius, an arc of the same length is cut by only two.

sity due to a flat-plate source diverge only near the plate edges (Fig. 1.2). The field due to an infinite plate, known in gravity work as the *Bouguer plate* (Section 2.5.1) and in radioactivity surveys as a source with '$2\pi$ geometry', does not vary with distance. In the real world, $2\pi$ geometry is approximately achieved when the detector is only a short distance above an extended source and a long way from its edges.

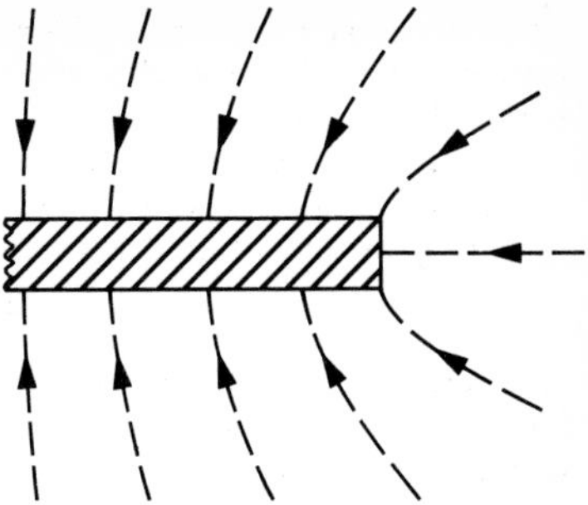

### 1.1.4 Dipoles

A dipole consists of equal-strength positive and negative point-sources a very small distance apart. Magnetic sources are fundamentally dipolar,

**Fig. 1.2** Lines of force from a semi-infinite slab. The lines of force diverge appreciably only near to the edge of the slab. This implies that elsewhere the field strength will decrease negligibly with distance.

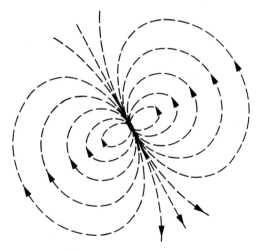

**Fig. 1.3** The dipole field.

3

while in some electrical surveys electrodes are used in approximately dipolar pairs.

The intensity of a dipole field decreases as the inverse cube of distance, and the direction changes with azimuth (Fig. 1.3). The intensity at a point along the dipole axis is double the intensity at the same distance along a line at right angles.

### 1.1.5 Vector addition

The effects of multiple sources are determined by vector addition of the individual fields. This form of addition (Fig. 1.4) must therefore be used to estimate the effect of a local source on a regional background. If the background is much larger, the resultant field will be approximately equal to the sum of the background field and the component of the anomalous field parallel to it. The direction of the resultant will differ slightly from that of the background but this effect is usually ignored.

If anomalies and background fields are similar in strength, there is no simple relationship between the magnitude of an anomaly and the magnitude of the resultant. The anomaly component in any direction

(usually the vertical) can be accurately determined by taking all measurements in that direction but this may not always be practicable.

The fields due to multiple sources are not necessarily those that would have existed had the various sources been present alone. A strong magnetic field from one body can affect the magnetisation in another, while the interactions between fields and conducting bodies in electrical and electromagnetic surveys can be very complex.

## 1.2 Geophysical equipment

Geophysical instruments vary widely in size and complexity but all are attempts to allow physical measurements of the sort commonly made in laboratories to be made at temporary sites under sometimes hostile conditions. They should be economical in power use, portable, rugged, reliable and simple. These criteria are satisfied with varying degrees of success by the commercial equipment currently available.

### 1.2.1 Choosing geophysical instruments

Few instrument designers can have

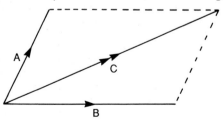

**Fig. 1.4** Vector addition by the parallelogram rule. Fields represented in magnitude and direction by lines A and B combine to give the resultant field C.

tried using their own products for long periods in the field and operator comfort seldom seems to have been considered. Moreover, although many real improvements have been made in the last twenty years, design features have been introduced during the same period, for no obvious reasons, which have actually made fieldwork more difficult. The proton magnetometer staff, discussed below, is a case in point.

If different instruments can, in principle, do the same job to the same standards, practical considerations become paramount. Some of the factors which need to be considered are listed below.

*Serviceability:* Is the manual comprehensive and comprehensible? Is a breakdown likely to be repairable in the field? Are there facilities for repairing major failures in the country of use or would the instrument have to be sent overseas, risking long delays in customs? Reliability is vital but many manufacturers seem to rely on their customers to evaluate prototypes.

*Power supplies:* If dry batteries are used, will they be easy to replace or impossible to find outside major cities? If the batteries are re-chargeable, can standard car batteries be used, either routinely or in emergencies? Are special cables or connectors needed to do this? How heavy are the batteries and how long will they continue to operate the instruments at the temperatures expected in the field?

*Data displays:* Are these clearly legible under all circumstances? A torch may be needed to read some in poor light, while some self-luminous displays are almost invisible in bright sunshine. Cathode-ray tubes exhaust power supplies very quickly.

*Comfort:* Is prolonged use likely to cripple the operator? Many instruments are designed to be used suspended on a strap passing across the back of the neck. This can be tiring enough under any circumstances but can cause real medical problems if the instrument has to be levelled by bracing it against the strap. Passing the strap over one shoulder and under the other arm may reduce the strain but not all instruments are convenient to use if carried in this way.

*Convenience:* If the instrument is placed on the ground, will it stand upright? Is the cable then long enough to reach the sensor in its normal operating position? If the sensor is mounted on a tripod or pole, is this strong enough? The traditional magnetometer poles, which simply screwed together and ended in a spiked section which could be stuck into soft ground, have now been largely replaced by unspiked, hinged rods which are more awkward to stow away, much more fragile (the hinges can twist and break), can only be used if fully extended and must be held or supported at all times.

*Fieldworthiness:* Are the control knobs and connectors protected from accidental impact? Is the casing truly waterproof? Does protection from damp grass depend on the instrument being laid down in a certain way? Are there depressions where moisture will collect and

5

then inevitably seep inside?

*Automation:* Has the manufacturer gone microprocessor mad? Switches have almost vanished from some modern instruments, on which every control operation has to be 'entered' via a keyboard. This may reduce problems from electrical 'spikes' generated by dirty contacts but, because the settings are not permanently visible, unsuitable values can easily be used in error.

## 1.2.2 Cables

Almost all geophysical work involves cables, which may be short, linking instruments to sensors or batteries, or hundreds of metres long. Efficient cable-handling is an absolute necessity.

Long cables always tend to become tangled, often because of well-intentioned attempts to make neat coils using hand and elbow. Coiling into a figure-of-eight rather than a simple loop is preferable, but even so it takes an expert to construct a coil from which cable can be run freely once it has been removed from the arm. On the other hand, a seemingly chaotic heap of wire spread loosely on the ground can be quite trouble-free. The basic rule is that cable must be fed to and from such heaps in opposite directions. Any attempt to pull cable from the bottom of the pile will certainly lead to disaster.

Cable heaps are unlikely to cause the permanent kinks which are often a feature of neat and tidy coils. Kinks can sometimes be removed by allowing the cable to hang freely and untwist naturally, but suitable places to do this with 100-metre lengths are rare.

Heaps can be made portable by feeding cables into open boxes, and on many seismic surveys the shot-firers carry their firing cables in this way in old gelignite boxes. Ideally, however, cables should always be wound directly on and off drums, although even then problems can occur. If cable is unwound by pulling the far end, the drum will not stop simply because the pull stops, and a free-running drum is an effective but untidy knitting machine. A drum mounted on a back-pack should be reversible so that it can be carried across the chest and wound or unwound from a standing position, and should have an effective brake. Many drums sold with geophysical instruments combine total impracticality with inordinate expense and home-made versions may be far better.

Problems with electrical interference (coupling or 'cross-talk') between cables are discussed in the sections dealing with methods where they are likely to be important. Mention must also be made of the almost hypnotic influence geophysical lines exert on livestock. Cattle have been known to desert lush pastures in favour of midnight treks through hedges and across ditches in search of juicy cables. Not only will the survey be delayed but a valuable animal may be killed by biting through insulation around a live conductor. Constant vigilance is the only solution.

## 1.2.3 Connections

Crocodile clips are usually adequate for electrical connections between single conductors. Multi-conductor contacts are made via heavy-duty plugs and sockets which are usually the weakest links in the entire field system. Faults are often caused by dirt which increases wear on the contacts.

Plugs are necessarily of types which clamp on to the cable, since tension will otherwise be borne by the weak soldered connections to the individual pins. However, cables tend to flex repeatedly just beyond the clamping point and wires may break within their insulated sleeving. Any break in a cable, or a broken or 'dry' joint, means work with a soldering iron. This is never easy when connector pins are already clotted with old solder, and is especially difficult if many wires crowd into a single plug. It is well worth reducing problems by following a few simple rules:

Plugs should always be carried, never dragged along the ground: two hands should always be used, one on the cable to take the strain of any sudden tug, the other to support the plug itself. The rate at which cable is reeled-in should never exceed a comfortable walking pace; especial care is needed when the last two or three metres are being wound on to a drum. Drums should have clips or sockets where plugs can be secured when not in use.

Plugs should be placed on the ground very gently and as seldom as possible: they must be protected from grit and moisture; socket units in particular are almost impossible to clean. Small plastic bags or 'cling-film' can be put to good use here.

## 1.2.4 Geophysics in the rain

Geophysicists generally need better protection from the elements than do geologists. The latter are able to keep moving and seek shelter quickly, but a geophysicist, huddled over his instruments, is a sitting target for rain, hail, snow and dust, as well as mosquitoes, snakes and dogs. His most useful piece of field clothing is often a large waterproof cape which he can not only wrap around himself but into which he can retreat like a tortoise into a shell, covering his instruments as well as himself (Fig. 1.5).

Electrical methods generally do not work in the rain, and heavy rain can be a source of seismic noise. Other types of survey can continue; most geophysical instruments are supposed to be waterproof and a few actually are. Unless dry weather can be guaranteed, a field party should be plentifully supplied with plastic bags and sheeting to protect instruments, and paper towels for drying them. Large transparent plastic bags can often be used to enclose instruments completely even while they are being used, but even then condensation may cause drift and erratic behaviour. Silica gel within instrument housings can be useful in absorbing minor traces of moisture, but cannot cope with large amounts, and a portable hair-drier at base camp may prove invaluable.

**Fig. 1.5** The geophysical cape in action. Magnetometer and observer are both dry, with only the sensor bottle exposed to the elements.

### 1.2.5 A geophysical toolkit

Regardless of the nature of a geophysical survey, similar tools are likely to be needed. A field toolkit needs to include the following:

- Long-nose pliers (the longer and thinner the better)
- Slot-head screwdrivers (one very fine, one normal)
- Philips screwdriver
- Allen keys (metric and imperial)
- Scalpel (light, expendable types are best)
- Wire cutters/strippers
- Spray-on electrical contact cleaner
- Fine-point 12 V soldering iron
- Solder and 'Solder-sucker'
- Multimeter (mainly for continuity and battery checks, so small size and durability are more important than high sensitivity)
- Torch (preferably of a type that will stand unsupported and double as a table lamp)
- Hand lens
- Insulating tape, preferably self-amalgamating
- Strong epoxy glue/'super-glue'
- Silicone grease
- Waterproof sealing compound
- Spare insulated and bare wire
- Spare insulating sleeving
- Kitchen cloths and paper towels
- Plastic bags and 'cling-film'

A comprehensive first-aid kit is equally important.

## 1.3 Geophysical data

Geophysical data are often gathered along straight traverses pegged at regular intervals, but readings may also be made at points distributed more or less randomly over a survey area. Even with random coverage, only natural-field measurements can be of truly 'point' data, since detectors are always placed some distance from artificial sources. Readings obtained at the same 'points' with different orientations will almost certainly be different.

## 1.3.1  Station numbering

Systems of station numbering should be logical and consistent. Where data are collected along traverses, numbers should define positions in relation to the traverse grid. Infilling between traverse stations 3 and 4 with stations $3\frac{1}{4}$, $3\frac{1}{2}$ and $3\frac{3}{4}$ is clumsy and may present typists with problems, whereas inserting a station 325E between 300E and 350E on Line 100N is easy and unambiguous. The fashion for labelling such a station $300+25$E has no discernible advantages and uses up the plus sign which may, with digital recording, have to stand in for N or E. It is worth defining the grid origin so that S or W stations do not occur, and this may be essential with microprocessor instruments which cannot cope with negatives.

If transmitters, receivers and/or electrodes are laid out along straight lines and the whole array can be reversed without changing the reading, the mid-point position should be recorded. Special notations are needed with asymmetric arrays, and indeed the increased likelihood of positioning error is a good reason for avoiding asymmetry. Great care is needed in recording relative positions of sources and detectors in seismic work.

Stations scattered randomly through an area are best numbered sequentially. Positions can be recorded in the field by pricking through the field maps or air-photos and labelling the reverse sides.

Recording grid references on the spot may seem desirable but is not generally so in practice. Mistakes are too easily made and valuable time may be lost while co-ordinates are measured. Geophysical observers are usually in more of a hurry than are field geologists, since their work may involve instruments which are on hire at high daily rates, draw power from batteries at frightening speed or are subject to drift.

If there are several observers, numbers can easily be accidentally duplicated. All field books and field sheets should record the name of the observer. The interpreter or data processor will need to know who to look for when things go wrong.

## 1.3.2  Recording results

Geophysical survey results are primarily numerical and must be recorded with even greater care than observations of field geology. Words may sometimes be difficult to read, but can usually be deciphered eventually; a set of numbers, on the other hand, may be wholly illegible or, even worse, may be misread.

Numbers may, of course, not only be misread but mis-written. The circumstances under which data are recorded in the field are varied but seldom ideal; observers are usually either too hot, too cold, too wet or too thirsty. Under such conditions, they may well delete correct results and replace them with incorrect ones, in moments of confusion or temporary dyslexia. Precise reporting standards must be enforced and strict routines must be followed if errors are to be minimized.

Data on geophysical field sheets should never be erased. Corrections should be made by crossing out the incorrect items, preserving their legibility, and writing the correct values alongside. Something may then be salvaged even if both the original and the correction are wrong.

Loss of geophysical data tends to be final; some of the material in a geological notebook might be remembered and re-recorded, but not strings of numbers. Duplication is therefore essential. This may be done in the field, using duplicating sheets or carbon paper, or by transcribing the results each evening. Whichever method is used, originals and duplicates must be separated immediately and stored separately thereafter. Duplication is useless if copies are stored together and lost together.

Non-numeric information is also important. Observers are uniquely placed to note and comment on a multitude of topographical, geo-logical, cultural and climatic factors which may affect the geophysical results. If they fail to do so, erroneous interpretations may be made. Geophysical field workers also have a responsibility to their geo-logical colleagues to pass on information of interest about places that only they may visit. They should at least be willing to record dips and strikes and return with rock samples.

### 1.3.3 Accuracy, sensitivity, precision

Accuracy and sensitivity must be carefully distinguished. A standard gravity meter, for example, is sensitive to field changes of one-tenth of a gravity unit, but an equivalent level of accuracy will be achieved only if readings are carefully made and drift corrections correctly applied. Accuracy is thus limited but not determined by instrumental sensitivity.

Accuracy and sensitivity must both be distinguished from precision, which is related merely to the numerical presentation and should always be appropriate to accuracy (Example 1.1). Not only does greater precision waste effort but false conclusions may be drawn if high accuracy is implied.

Geophysical measurements can sometimes be made to an accuracy greater than is needed or even usable by the interpreters. Even so, the highest possible accuracy should always be sought, as later advances may allow more effective analysis. Reading the instrument twice at each occupation of a station reduces the incidence of major errors.

**Example 1.1**

---

Gravity reading = 858.3 scale units (s.u.)
Calibration constant = 1.0245 g.u./s.u. (see Section 2.1)
Converted reading = 879.32835 gravity units (g.u.)
But reading accuracy is only 0.1 g.u. (approximately),
and therefore:
Converted reading = 879.3 g.u.

(Note that the four decimal-place precision is necessary in the calibration constant: 858.3 multiplied by 0.0001 is equal to almost 0.1 g.u.)

---

### 1.3.4  Drift

Geophysical instruments will not usually record the same results if read repeatedly at the same place. The 'drift' may be due to changes in background field but often to changes in the instrument itself. Correcting instrumental drift is an essential first stage in data analysis.

Drift is often related to temperature and is unlikely to be linear between two readings taken in the relative cool at the beginning and end of the day if temperatures are ten or twenty degrees higher at noon. Survey 'loops' may therefore have to be limited to periods of only one or two hours.

Changes in background field are sometimes allowed for in drift corrections, but in most cases the variation can either be monitored directly (as in magnetic surveys) or calculated (as in gravity work). Where such alternatives exist, it is preferable they be used, since poor instrument performance may otherwise be overlooked. Drift calculations should be completed before a field crew leaves a survey area.

Drift corrections are most critical in gravity surveys and are discussed further in Section 2.4.5.

### 1.3.5  Signals and noise

To a geophysicist, *signal* is the effect that he is interested in measuring and *noise* is anything he also measures but which he considers to contain no useful information. One man's signal may well be another's noise; the magnetic effect of a buried pipe is a nuisance to anyone trying to draw geological conclusions from magnetic data but may be invaluable to a site developer who needs to know where the pipe runs. Much geophysical field practice is dictated by the need to improve signal-to-noise ratios.

The statistics of random noise are important in seismic and radiometric work and with induced polarization (IP) receivers. Adding together $N$ statistically long random series, each of average amplitude $A$, will produce a random series with amplitude $A\sqrt{N}$. Since $N$ identical signals of average amplitude $A$

treated in the same way produce a signal of amplitude $A \times N$, adding $N$ signals with random noise components can improve signal-to-noise ratios by a factor equal to $\sqrt{N}$.

### 1.3.6 Variance and standard deviation

Statistical methods play a rather small part in field geophysics, although they are fundamental to the processing of seismic data. Only reading errors, radioactive count rates and seismic noise can be expected to obey statistical rules. Background variations such as short-term changes in magnetic field must be precisely measured.

Random variations often follow a Normal (Gaussian) Distribution with a bell-shaped probability curve (Fig. 1.6). Such a distribution can be characterized by a mean, equal to the sum of all the values, divided by the total number of values, and a variance or its square-root, the standard deviation (SD). Variance is defined as:

$$V = \frac{1}{N-1} \sum_{n=1}^{N} x_n^2$$

where $x_n$ is the difference between the $n$th value and the mean of the set of $N$ values.

About two-thirds of the readings in a Normal Distribution lie within

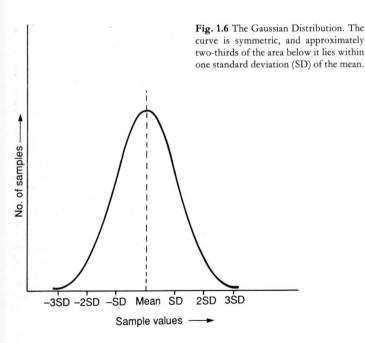

**Fig. 1.6** The Gaussian Distribution. The curve is symmetric, and approximately two-thirds of the area below it lies within one standard deviation (SD) of the mean.

No. of samples ——►

−3SD −2SD −SD Mean SD 2SD 3SD

Sample values ——►

1 SD of the mean, and less than 0.3% differ from it by more than 3 SDs. The SD is popular with contractors estimating survey reliability, since a small value can effectively conceal several major errors. Normal Distributions are often assumed rather than proven.

### 1.3.7 Profiles and contours

If geophysical data are collected along traverse lines, the results can be presented in profile form, as in Fig. 1.7. The horizontal scale is always distance; the vertical scale is of the quantity being measured. It is generally possible to plot profiles in the field, or at least each evening, as work progresses. Such plots have a vital place in quality control of fieldwork.

Traverse lines drawn on topographic maps can be used as baselines for profile plots. Such presentations are particularly helpful in identifying anomalies due to man-made features, since correlations with roads and field boundaries are obvious. If profiles along a number of different traverses are presented in this way on the one map they are said to be *stacked*, although elsewhere in geophysics the same word is used to describe the combination of multiple data sets to give a single output.

Contour maps tend to be drawn in the field only if the strike of some feature must be defined so that infill work can be planned. Information is lost in contouring because it is not generally possible to choose a contour interval which faithfully

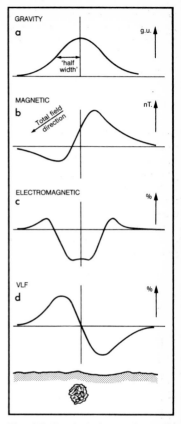

**Fig. 1.7** Geophysical anomaly profiles across a pyrrhotite-rich sulphide orebody. The amplitude of the gravity anomaly (*a*) might be a few g.u. and of the magnetic anomaly (*b*) a few hundred nanoTesla (nT). The electromagnetic anomaly (*c*) is for a system in which transmitter and receiver are both small coils with vertical axes, the VLF anomaly (*d*) for a system which measures the dip of the magnetic component of a very low frequency radiowave. Neither of these latter two anomalies is likely to have an amplitude of more than 25%.

records all the features of the original data. Also, noise is introduced because contour lines are drawn across areas between traverses where no data are available.

A special form of 'cross-sectional' contour map, the *pseudo-section*, is described in Section 7.4.2.

## 1.3.8 Anomalies

Only rarely is a single geophysical observation significant. Many readings are needed and a background level must be determined before geological interpretation can begin. Interpreters tend to concentrate on 'anomalies', where there are differences from a constant or smoothly varying background.

Geophysical anomalies take many forms. A massive sulphide deposit containing pyrrhotite would be dense, magnetic and electrically conductive. Typical anomaly profiles obtained over such a body in various types of geophysical survey are shown in Fig. 1.7. A wide variety of possible contour patterns correspond to these differently shaped profiles.

Background fields also vary and may, at different scales, be regarded as anomalous. A 'mineralization' gravity anomaly, for example, may lie on a broader high, due to a mass of basic rock. Separation of observations into patterns of regionals and residuals is an important part of geophysical data processing and even in the field it may be necessary to estimate background so that the significance of local anomalies can be assessed.

The existence of an anomaly indicates a difference between the real world and some idealized model, and in gravity work the terms *free-air*, *Bouguer* and *isostatic anomaly* are used to denote derived quantities which record differences from gross Earth models. These 'anomalies' are sometimes almost constant within small survey areas.

## 1.3.9 Wavelengths and half-widths

Geophysical anomalies in profile resemble transient waves but vary in space rather than time. In describing them the terms *frequency content* and *frequency* are often loosely used, although *wavenumber* (the number of cycles in unit distance) is pedantically correct. *Wavelength* may be quite properly used of a spatially varying quantity, but where geophysical anomalies are concerned the usage is imprecise; an anomaly described as having 'a wavelength', would be resolved by Fourier analysis into a number of components of different wavelengths.

A more precisely defined quantity is the *half-width*, which is half the distance between the points at which the amplitude has fallen to half the anomaly maximum (cf. Fig. 1.7a). This is roughly equal to a quarter of the wavelength of the dominant sinusoidal component, but has the advantage of being directly measurable on field data.

## 1.4 Bases and base networks

*Bases* are important in gravity and magnetic surveys, and in some electrical and radiometric work. They may be:

1  Repeat stations which mark the starts and ends of sequences of readings and which are used to determine drift.
2  Reference points at which the value of the field being measured has already been established by previous work.
3  Points at which regular measurements of background are made whilst field readings are taken elsewhere.

A single base may fulfil more than one of these functions. The reliability of survey results and the ease with which later work can be related to them often depends on the base stations.

### 1.4.1 Base station principles

There is no absolute reason why any of the three types of base should coincide, but surveys tend to be simpler and fewer errors occur if some rules are followed. It is desirable that every 'drift base' also be a 'reference base' and if, as is usually the case, there are too few existing reference points for efficient working, the first step in a survey should be to establish an adequate base network.

It is not essential that a diurnal base be part of this network and it may actually be inconvenient for it to be so. However, if a diurnal monitor is used, the day's work will normally be begun by setting it up and end with its removal. It is good practice to read the field instruments at a point at, or near, the monitor position on these occasions, noting any differences between base and field readings.

### 1.4.2 ABAB ties

Bases are normally linked together using ABAB ties. A reading is made at Base A and the instrument is then taken as quickly as possible to Base B. Repeat readings are then made at A and B. The last reading at B may also be the first in a similar set linking B to a Base C, in a process known as *forward-looping*.

Each set of four readings provides two estimates of the difference in field strength between the two bases, and if these do not agree within the limits of instrumental accuracy ($\pm 1$ nT in Fig. 1.8), further links should be made. Differences should be calculated in the field so that extra links may be added if needed. The times between readings should be short so that drift and sometimes also diurnal variation can be assumed linear.

### 1.4.3 Base networks

Modern geophysical instruments are very accurate and quite easy to read, so that the error in any single esti-

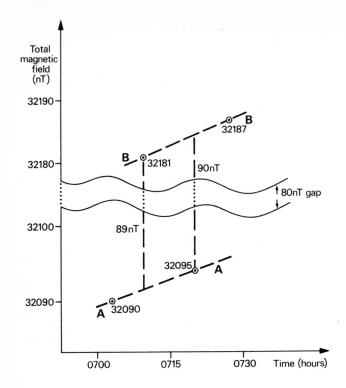

**Fig. 1.8** ABAB tie between bases in a magnetic survey using a magnetometer with 1 nT sensitivity. A difference estimate of 89 nT between the two stations would be accepted. Note that the scale of the plot must be appropriate to instrument sensitivity and that it may be necessary to 'remove' some of the range from the graph to allow points to be plotted with sufficient precision.

mate should be trivial. However, a final value obtained at the end of an extended series of links could include quite large accumulated errors.

The integrity of a system of bases can be assured if they form part of a network in which each base is linked to at least two others. Misclosures are calculated by summing differences around each loop, with due regard to sign; these are then reduced to zero by making the smallest possible adjustments to individual differences. The network in Fig. 1.9 is

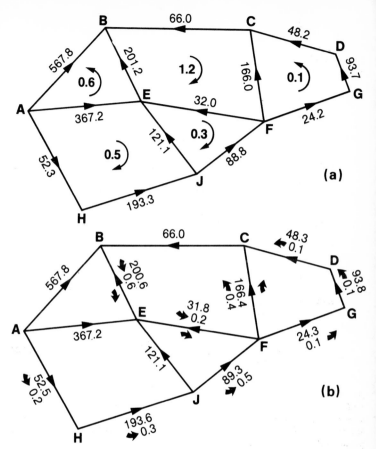

**Fig. 1.9** Network adjustment. (*a*) The 1.2 g.u. misclosure in BCFE suggests a large error in either the 'unsupported' link BC or in BE, the only link shared with another loop with large misclosure. (*b*) Adjustments made on the assumption that BC was verified but that no other checks could be made. The large adjustments needed for BE, JF and CF are beyond the levels normally considered acceptable in modern gravity surveys.

sufficiently simple to be adjusted by inspection. A more complicated network could be adjusted by computer, using a least-squares or other criterion, but this is generally unnecessary in small-scale surveys.

## 1.4.4 Siting base stations

It is obviously important that bases be adequately described and, where possible, permanently marked, so that extensions or fill-ins can be linked to previous work by exact re-occupations. Unfortunately, conventional concrete or steel markers tend to be quickly obliterated, either deliberately or accidentally.

A solution to the problem of the impermanence of 'permanent' markers is to describe some of the station locations in terms of existing features which are unlikely to be moved. In any survey area there are points distinctive in their own right because of the presence of man-made or natural features. Describing and sketching such points is the best way to preserve information for the future; good sketches are certainly better than most photographs.

Finding permanent features may sometimes be difficult, and maintaining gravity bases at international airports is a continual problem. Geodetic survey markers are usually secure but are often in isolated and exposed locations. Statues, memorials and historic or religious buildings often provide sites which are not only quiet and permanent but also offer some degree of shelter from sun, wind and rain.

Individual geophysical methods have their own unique base-station requirements which are discussed in the appropriate chapters.

# 2

# Gravity Method

Differences in rock density produce small changes in gravity field which can be measured using portable instruments known as *gravity meters* or *gravimeters*.

## 2.1 Physical basis of the gravity method

A gravity field is equivalent to an acceleration, for which the S.I. unit is the metre·sec$^{-2}$. This is too large for geophysical work and the gravity unit (g.u.), $10^{-6}$ m·sec$^{-2}$, is generally used. The c.g.s. unit, the milligal, equal to 10 g.u., is still very popular.

### 2.1.1 Gravity field of the Earth

The Earth's gravity field is almost equal to that of a sphere with the same average radius and total mass, but increases slightly towards the poles. The relationship between normal sea-level gravity and latitude ($L$) is described by the *International Gravity Formula*, adopted in 1967:

$$g_{norm} = \begin{aligned} & 9\,780\,318.5 \\ & + 51629.27\,\sin^2 L \\ & + 229.5\,\sin^4 L \end{aligned}$$

This formula, which replaced a 1930 version with slightly different constants, is compatible with a network of international base stations known as IGSN71: 9 780 318.5 is the theoretical sea-level gravity, in g.u., at the Equator. A still more recent (1980) version with constants 9780327, 52790.4 and 232.7 is only slowly finding acceptance. The difference between polar and equatorial fields is about 50 000 g.u., the rate of change reaching a maximum of about 8 g.u. per kilometre north or south at 45° latitude (Fig. 2.1).

Gravity anomalies due to geological sources are small compared to the Earth's main field. A major sedimentary basin may reduce the gravity field by more than 1000 g.u., but massive ore bodies, which are also common targets, produce anomalies of only a few g.u. Topographic effects may be much larger; there should be a difference of nearly 20 000 g.u. between gravity at the

19

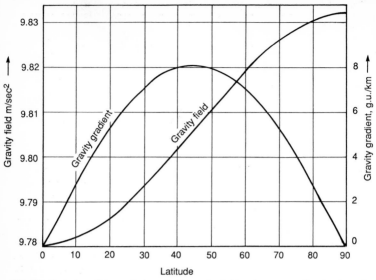

**Fig. 2.1** Variation of theoretical sea-level gravity field and corresponding horizontal gradient with latitude. Note that there is no East–West gradient.

summit of Mt. Everest and at sea-level due to elevation difference alone. For geological purposes, gravity changes must often be measured to an accuracy of 0.1 g.u. (approximately one-hundred milli-onth of the Earth's field), the sensitivity of virtually all modern gravity meters.

### 2.1.2 Rock density

Most crustal rocks have densities of between 2000 and 2800 kg·m$^{-3}$ (Water density = 1000 kg·m$^{-3}$). In the early days of gravity work a density of 2670 kg·m$^{-3}$ was adopted as standard for the upper crust and

is still widely used in modelling and in calculating elevation corrections for standardized gravity maps. Density ranges for some common rocks are shown in Table 2.1.

### 2.2 Gravity meters

For the past fifty years the vast majority of gravity measurements have been made using meters with astatic spring systems, and this seems likely to be the case for the foreseeable future. Such meters measure gravity differences, not absolute field strengths, and many of the problems encountered in gravity surveys stem from this fact.

**Table 2.1** Densities of common rocks and ore minerals (kg·m$^{-3}$)

*Common rocks*

| | | | |
|---|---|---|---|
| Dry sand | 1400–1650 | Serpentinite | 2500–2600 |
| Wet sand | 1950–2050 | Gneiss | 2650–2750 |
| Coal | 1200–1500 | Granite | 2500–2700 |
| Chalk | 1900–2100 | Dolerite | 2500–3100 |
| Salt | 2100–2400 | Basalt | 2700–3100 |
| Limestone | 2600–2700 | Gabbro | 2700–3300 |
| Quartzite | 2600–2700 | Peridotite | 3100–3400 |

*Ore minerals*

| | | | |
|---|---|---|---|
| Sphalerite | 3800–4200 | Galena | 7300–7700 |
| Chalcopyrite | 4100–4300 | Chromite | 4500–4800 |
| Pyrrhotite | 4400–4700 | Hematite | 5000–5200 |
| Pyrite | 4900–5200 | Magnetite | 5100–5300 |

### 2.2.1 Astatic spring systems

Astatic systems use 'zero-length' main springs, in which the tension is proportional to the actual length. With the geometry shown in Fig. 2.2 and for one particular value of gravity field, the spring will support the balance arm in any position.

In stronger fields, an auxiliary spring can be used to support the increase in weight, which will be equal to the product of the mass and the increase in gravity field. This spring can be quite weak; to use an expression common in other areas, the zero-length spring 'backs-off' a constant weight so that the measuring spring can respond to small changes in gravity field.

### 2.2.2 Commercial gravity meters

None of the common gravity meters in Fig. 2.3 uses exactly the system of Fig. 2.2. The Worden and Sodin have two auxiliary springs, one for fine and one for coarse adjustments, attached to balance arms of more complicated design. The LaCoste has no auxiliary but the main spring is not quite zero length. Some form of temperature compensation is also usually included.

Spring systems are mechanical and therefore subject to drift. Short-period drift is largely due to temperature changes which affect the elastic constants of the springs. There is also longer-term 'creep', by which the springs slowly extend with time. Drift must be monitored by repeated readings at base stations.

Readings are obtained by positioning a movable line or pointer, viewed through a magnifying eyepiece, at a fixed point on a graduated scale. The pointer is moved by rotating a calibrated dial. Because

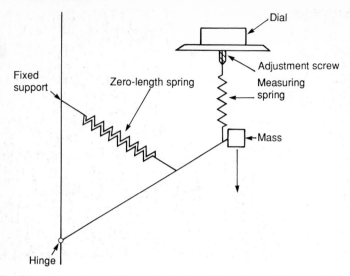

**Fig. 2.2** Astatic gravimeter. The tension in the zero-length spring is proportional to its length and not, as with most springs, merely to its extension beyond some unstressed length. Measurements are made by restoring the mass to a standard position, using the measuring spring and dial.

alignment of pointer and scale is rather subjective, all measurements in a given 'loop' should be made by the same observer. The subjective element is then eliminated when differences are calculated. The reading, usually a combination of a value read from the dial and an associated value displayed on a mechanical counter, is converted to gravity units using a calibration constant specific to the instrument.

Because astatic systems are asymmetric, the effect of a levelling error depends on the direction of tilt. A slight off-levelling at right-angles to the balance arm gives a reading of gravity field multiplied by the cosine of the tilt angle (an error of about 0.15 g.u. for a tilt of 0.01°). If the tilt is in the plane of movement, reading sensitivity (the amount the pointer moves for a given rotation of the dial) is also affected.

### 2.2.3 Quartz astatic meters

Worden and Sodin meters have springs of fused quartz, enclosed in vacuum chambers which provide a high degree of thermal insulation. Some also have electrical thermostats. The meters are pressure-sensitive, the effect amounting to about 1 g.u. for a 500 metre elevation change. No quartz meter can be

**Fig. 2.3** LaCoste 'G' (geodetic) (left), Worden 'Student' (centre) and Sodin 420-T (right) gravimeters.

clamped for transit, so the spring systems are vulnerable; if a meter suffers sharp sideways acceleration or is tilted, even gently, through more than about 45°, the springs may become tangled, requiring re-alignment by the manufacturer.

Quartz meters have limited ranges, generally of between 500 and 2000 g.u., on the direct-reading scale, and must be re-set if a limit is reached. In some models the re-set can be made with a second dial calibrated to a lesser degree of accuracy, but in others an uncalibrated adjustment screw is used. It is always advisable to allow several hours for the system to 'settle' after re-set.

The levels of Sodin meters are deep within the instrument and can only be seen if illuminated. They are thus shielded from the direct rays of the sun, which in other meters can cause levelling errors due to uneven heating of the fluid. However, re-checking of the levels during reading cannot be the almost automatic process it is when they are set in the top surface of the meter case, and it seems probable that more readings are taken off-level with Sodins than with other meters.

Quartz meters are often quite difficult to read and level, and it takes an experienced and conscientious observer to attain the theoretical reading accuracy of 0.1 g.u.

### 2.2.4 Steel astatic meters

The LaCoste–Romberg is roughly

23

the same size and weight as the Worden or Sodin, but very different in appearance, being enclosed in an almost cubical box. Its effective weight of about 5 kg is almost doubled by the 12-V battery which, because the steel spring has a high thermal conductivity and cannot be effectively insulated, must be permanently connected to the thermostat element. Until two or three hours after reaching its operating temperature the instrument drifts so severely as to be unusable. Drift is then very low, and can even be extrapolated linearly across intervals during which the meter has been 'off-heat'. However, discontinuous 'tares' of perhaps several g.u. can occur.

The LaCoste spring can be clamped, and when clamped is reputedly able to survive any shock which does not fracture the outer casing. The optical system is clearer than that of most quartz meters and the spring is less affected by vibration. Even quite inexperienced observers have no trouble in attaining 0.1 g.u. accuracy, particularly if aided by the optional, and expensive, electronic readout.

Another major difference between the LaCoste 'G' (geodetic) meter and all others is that a single long measuring screw is used to give readings world-wide without re-set. Calibration factors vary slightly over the range, being tabulated for 1000 g.u. intervals.

A LaCoste thus has considerable advantages, but costs about twice as much as quartz equivalents and needs heavy batteries to power the heater. A single charge lasts one or two days, depending on thermostat setting and external temperature, and some form of charger is needed in the field.

## 2.2.5 Reading a gravity meter

Gravity meters are normally read on concave 'dishes' supported by three short stubs to which longer legs can be attached. The stubs are usually used alone, pressed firmly but not too deeply into the ground; the undersurface of the dish should not touch the ground since this would provide a fourth support point and allow 'rocking' back and forth.

Thick grass can 'ball-up' under a dish and may have to be removed before a reading can be taken. The extension legs may be used where such conditions are common but readings will then take longer, the dish itself may have to be levelled (some incorporate a 'bull's-eye' bubble) and the height above ground will have to be measured.

The meters themselves rest on three adjustable, screw-threaded 'feet' and are levelled using two horizontal spirit-levels (Fig. 2.4), initially by being moved around the dish until both level bubbles are 'floating'. The temptation to hurry this stage and use the footscrews almost immediately should be resisted.

It will usually be found that one of the levels (probably the cross-level, at right angles to the plane of movement of the balance arm) is set parallel to a line joining two of the

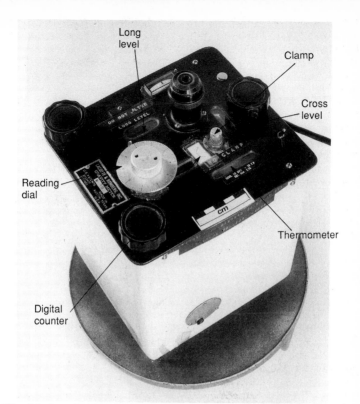

Long level

Clamp

Cross level

Reading dial

Thermometer

Digital counter

**Fig. 2.4** Controls of the LaCoste 'G' gravity meter. Note the two level bubbles at right angles, the clamp and the aluminium reading dial. The digital counter is behind the small window between clamp and dial. The thermometer, underneath the window in front of the clamp, monitors internal temperature and *must* record the pre-set operating temperature if the instrument is to be usable.

feet. Adjustments to the third foot then scarcely affect the bubble in this level. Some meters can rotate in their casings and level-bubbles and feet may become misaligned, but levelling is very much easier if any such slippage is corrected.

The quickest method of levelling is to centre the cross-level bubble, using one or both of the two screws which control it, and then use the third screw to set the long level. Experienced observers may use two screws simultaneously, but the ability to do this efficiently comes only with practice.

Once a meter is level, a reading can be obtained by setting the pointer in the correct position. It is vital that the level bubbles are checked whilst the dial is being adjusted, and especially immediately after a supposedly satisfactory pointer position has been reached. Natural surfaces tend to subside slowly under the weight of the observer, off-levelling the meter. On snow or ice the levels have to be adjusted almost continuously as the dish melts its way down. The problem can be avoided by first bedding-down a dish-sized piece of plywood and setting the dish up on that.

All mechanical measuring systems suffer from 'whiplash', i.e. two readings will differ, even if taken within seconds of each other, if the final adjustments are made by opposite rotations of the reading dial. The only remedy is total consistency in the direction of the final adjustment.

Severe continuous vibration, as from nearby machinery or the roots of trees moved by the wind, can cause not merely difficulty in obtaining a reading but actual shifts in value. Also, even distant earthquakes can make the pointer swing slowly from side to side across the field of view. Fortunately, this effect is rare in most parts of the world; if it is seen, survey work must be stopped until the disturbance is over.

## 2.2.6 Meter checks

A series of checks should be made each day before beginning routine survey work. The meter should first be 'shaken-down', by tapping it gently on the dish between readings until a constant value is recorded.

The levelling system also needs to be checked. Off-levelling a correctly adjusted cross-level will reduce the reading, regardless of the direction of offset. The meter should be set up and read normally and the cross-level should then be offset by equal amounts in first one direction and then the other. The pointer should move roughly the same distance in the same direction in each case; meters may be considered usable provided that the movements are at least in the same direction.

The long-level affects reading sensitivity, i.e. the distance the pointer will move for a given dial rotation. The manufacturer's handbook will list the optimum sensitivity and the actual value can be found by moving the dial a given amount and observing the pointer movement.

Instructions for re-setting levels will be found in the handbooks. After re-set, levels often 'settle-in' to their new positions over a period of days, during which time they must be re-checked with special care.

## 2.2.7 Meter calibration

The calibration constant which converts dial readings to gravity units is not affected by changes in reading sensitivity but does change slowly with time and must be checked regularly. This can be done by the manufacturers or by using calibration ranges of known gravity interval. Calibration ranges usually involve gravity changes of about 500 g.u., which is slightly less than the range of the most limited-range meters, and almost always make use of the rapid change of gravity field with elevation. A height change of about 250 metres is generally necessary, although in some cases local gravity gradients can also play a part. Travel time between top and bottom stations should normally be less than 15 minutes and the two stations should be well marked and described. A 'run' involves reading alternately at the two stations, with at least two occupations of each, giving two estimates of gravity difference (cf. Section 1.3.2). If these differ by more than 0.1 g.u., more 'legs' should be added.

Meters with separate fine and coarse adjustments can be checked over different sections of their fine ranges by slightly altering the coarse setting. Most meters need a little time to stabilize after coarse adjustment, but if this is allowed it may be possible to identify minor irregularities in a calibration curve. This cannot be done with LaCoste 'G' meters, since only one part of the curve can be monitored on any one calibration range. Different LaCostes may give results on the same range which differ consistently by a few tenths of a g.u.

Calibration factors are usually quoted by manufacturers in milligals, not g.u., per scale division and involve the rather arbitrary insertion of a decimal point somewhere in the reading. Any uncertainty can be resolved by remembering that the sensitivity of all modern meters (except LaCoste high-precision tidal meters) is such that the last figure in the reading approximately corresponds to tenths of a g.u.

## 2.3 Gravity reductions

In gravity surveys, more than in any other geophysical method, large and (in principle) calculable effects are produced by sources which are not of direct geological interest. These effects are removed by 'reductions' involving sequential calculation of a number of recognized anomaly values. In each case the sign of the reduction is opposite to that of the effect it is designed to remove. A positive effect is one which has a component in the direction of the Earth's main field.

### 2.3.1 Latitude correction

Latitude corrections are usually made by subtracting the 'normal' gravity, calculated from the International Gravity Formula, from the 'observed' gravity. For surveys not tied to the absolute reference system, local latitude corrections may be made by selecting an arbitrary base and calculating the theoretical local gradient, equal to

$$8.12 \sin 2L \text{ g.u./km}$$

where $L$ is the latitude angle.

### 2.3.2 Free-air correction

The remainder left after subtraction

of the normal from the observed gravity will be due in part to the height of the gravity station above the sea-level reference surface. Since an increase in height implies an increase in distance from the Earth's centre of mass, the effect is negative for stations above sea level, and the correction, termed the *free-air correction*, is positive, and for all practical purposes equal to 3.086 g.u./metre.

The quantity obtained after applying both the latitude and free-air corrections is the free-air anomaly.

### 2.3.3 Bouguer correction

Since topographic masses are irregularly distributed, their effects are difficult to calculate precisely and approximation is necessary. The simplest approach assumes that topography can be represented by a flat plate of uniform density which extends to infinity on all sides, with a thickness equal to the height of the gravity station above the reference surface. This approach was first suggested by Bouguer; the gravity field of a plate of thickness $h$ and density $\rho$ is:

$$b = 2\pi\rho Gh$$
$$b_{2670} = 1.119 \text{ g.u./metre}$$

The field due to the slab is positive and the correction is negative. Since it is only about one-third of the size of the free-air correction, the net effect of an increase in height is a reduction in field, and the net correction is positive and equal to about

2 g.u. per metre. Elevations must be known to 5 cm to make full use of meter sensitivity.

Because the Bouguer correction depends on an assumed density as well as a measured height, there is a fundamental difference between it and the free-air correction and combining the two into a single 'elevation' factor may be misleading.

### 2.3.4 Terrain corrections

In areas of high relief, detailed topographic corrections must be made. Although it would be possible to correct directly for the entire topography above the reference surface without first computing the Bouguer correction, it is mathematically simpler to calculate the Bouguer anomaly and then correct for deviations of topography from the Bouguer plate.

A peculiarity of the two-stage approach is that terrain corrections are always positive. The mountain mass (A) above the gravity station in Fig. 2.5 exerts an upward pull on the gravity meter, and the correction is obviously positive. The valley (B) lies in a region that the Bouguer calculation assumed to be filled with rock, the gravity pull of which would have had a downward component. The terrain correction must thus allow for over-correction by the Bouguer plate and is again positive.

Terrain corrections are extremely tedious. In manual work, a transparent *Hammer chart* is centred on the gravity station on the topographic

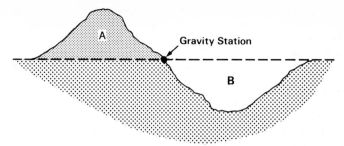

**Fig. 2.5** Terrain corrections. The corrections are concerned with deviations of the topography from a horizontal plane through the gravity station, not from the sea-level or other reference surface, and are always positive. A = mountain mass; B = valley.

map (Fig. 2.6) and the difference between the average height of the terrain and the station height is estimated for each compartment. The corresponding corrections are obtained from tables (see Appendix). Computer methods have been developed, but these require digitization of the terrain in some form or other and may be equally time-consuming.

Adding the terrain correction to the simple Bouguer anomaly produces a quantity often known as the *Extended Bouguer Anomaly*. Topographic densities are sometimes varied with geology in the calculations to still further minimize the effects of changes in station elevation.

## 2.4 Gravity surveys

A gravity survey is basically a very simple operation. Nonetheless, few are completed wholly without problems, and in some cases the outcome can only be described as disastrous. Most of the difficulties arise because

gravity meters measure only differences in gravity field and readings have to be inter-related by links to a common reference system.

### 2.4.1 Survey principles

An individual gravity survey 'loop' consists of a series of observations which begin and end with readings at the same point, the *drift base* (Section 1.4). The physical size of the loop is usually dictated by the need to monitor drift at roughly two-hour intervals and so will vary with the mode of transport being used. At least one station of the reference network must be occupied in the course of each loop and survey operations are simplified if every drift base is such a station.

In principle, a base network can be allowed to emerge gradually as the work proceeds. In practice, it is very much easier if the net is completed and adjusted at a very early stage, since absolute values at 'field' stations can then be obtained immediately readings are made at them. There is also much to be gained from the early overview of the whole survey area which is obtained while the network is being set up. Furthermore, there are practical advantages in establishing bases while not under the pressure to max-

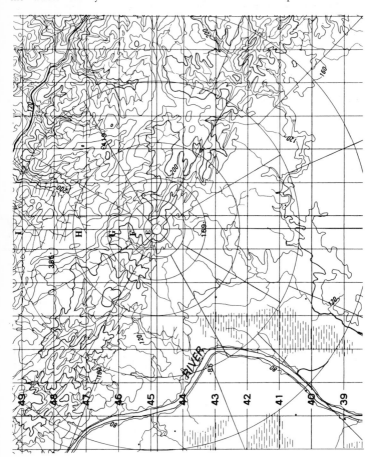

**Fig. 2.6** Hammer chart, showing Zones E to I, overlaid on a topographic map. The problems in estimating average heights in the various compartments are easily seen. The letters identifying the zones are difficult to see in this example, which shows the chart overlay in place on the map, but are in practice easily visible when the overlay is removed and viewed on its own.

imize the daily total of new stations which characterizes the routine production phase of most surveys.

A small survey may define local gravity anomalies in relation to an arbitrary base and without any tie to the absolute system. Problems will arise only when such a survey has to be linked to others or added to a national database. Since this nearly always happens eventually, use of a purely local reference is generally a false economy.

### 2.4.2 Base stations

The criteria used in selecting reference bases are quite different from those for normal stations. Provided that exact re-occupation is possible, large terrain effects can be tolerated. These may make it inadvisable to use the Bouguer anomaly interpretatively, in which case the elevation is not needed either. On the other hand, since the overall survey accuracy depends on repeated base readings, quiet environments and easy access are important. Traffic noise and other strong vibrations can invalidate base (or any other) readings.

The general principles outlined in Section 1.4 apply to gravity bases. Descriptions should be provided for the main survey bases, in the form of sketch plans which will allow re-occupation to within a few centimetres (Fig. 2.7). It is particularly important that the re-occupations are made at the same elevation as the original reading.

### 2.4.3 Station positioning

The sites of field stations also need to be chosen with care. Except in detailed surveys with stations at

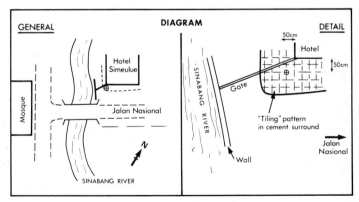

**Fig. 2.7** Gravity base station description sketch. Sketches at two different scales (as shown) are usually needed, together with a short written description, to ensure that the station can be reoccupied quickly and accurately.

fixed intervals along traverses, the observer in the field has some, and often considerable, freedom of choice. Furthermore, it is his responsibility to estimate terrain corrections within about 100 metres of the reading point, since features within this zone which are too small to be shown on any topographic map can be significant. The magnitude of near-zone terrain corrections must be appreciated and common sense must be used in selecting reading points where these will be small.

Local effects can be estimated using a truncated graticule; the example in Fig. 2.8 covers the Hammer Zones B and C only. Height differences of less than 0.3 metre in Zone B and 1.3 metres in Zone C can

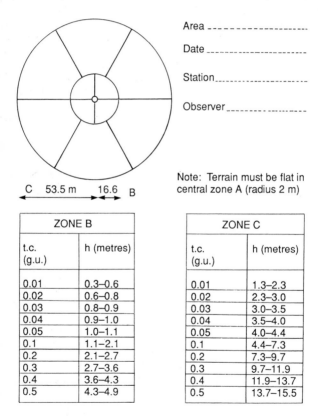

Area _ _ _ _ _ _ _ _ _ _ _ _ _ _ _ _ _ _

Date _ _ _ _ _ _ _ _ _ _ _ _ _ _ _ _ _

Station _ _ _ _ _ _ _ _ _ _ _ _ _ _ _ _

Observer _ _ _ _ _ _ _ _ _ _ _ _ _ _

Note: Terrain must be flat in central zone A (radius 2 m)

C    53.5 m    16.6    B

| ZONE B | |
|---|---|
| t.c. (g.u.) | h (metres) |
| 0.01 | 0.3–0.6 |
| 0.02 | 0.6–0.8 |
| 0.03 | 0.8–0.9 |
| 0.04 | 0.9–1.0 |
| 0.05 | 1.0–1.1 |
| 0.1 | 1.1–2.1 |
| 0.2 | 2.1–2.7 |
| 0.3 | 2.7–3.6 |
| 0.4 | 3.6–4.3 |
| 0.5 | 4.3–4.9 |

| ZONE C | |
|---|---|
| t.c. (g.u.) | h (metres) |
| 0.01 | 1.3–2.3 |
| 0.02 | 2.3–3.0 |
| 0.03 | 3.0–3.5 |
| 0.04 | 3.5–4.0 |
| 0.05 | 4.0–4.4 |
| 0.1 | 4.4–7.3 |
| 0.2 | 7.3–9.7 |
| 0.3 | 9.7–11.9 |
| 0.4 | 11.9–13.7 |
| 0.5 | 13.7–15.5 |

**Fig. 2.8** Field observers' chart for terrain corrections (t.c.), Hammer Zones B and C.

be ignored, as producing effects of less than 0.01 g.u. per compartment.

It can easily be shown that the effect of a normal survey vehicle is detectable only if the observer actually crawls underneath it, and most modern buildings produce similarly small effects. Old, thick-walled structures may need to be treated with more respect (Fig. 2.9). Sub-surface cavities, whether cellars, mineworkings or natural caverns, can produce anomalies amounting to several g.u. The gravity method is sometimes used in cavity detection but, where this is not the object of the survey, it is obviously important that such effects are eliminated from the gravity maps by careful station siting.

### 2.4.4 Tidal effects

Before meter drift can be estimated allowance must be made for *tidal effects*. These are background variations due to changes in the relative positions of the Earth, moon and sun, and follow linked 12 and 24-hour cycles, superimposed on a cycle related to the lunar month (Fig. 2.10). Swings are largest at new and full moons, when the Earth, moon and sun are in line and when changes of more than 0.5 g.u. may occur within an hour and total changes may exceed 2.5 g.u. The assumption of linearity made in correcting for drift may fail completely if tidal effects are not first removed.

Earth tides are entirely predict-

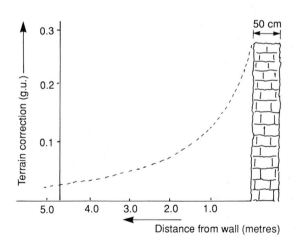

**Fig. 2.9** Effect of a 50 cm-thick stone wall on a gravity meter.

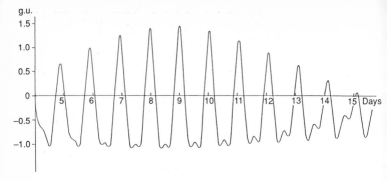

**Fig. 2.10** Tidal variation in gravity units (g.u.), 5th to 15th of January 1986.

able, at least at the 0.1 g.u. level required for gravity survey. Tables from which the corrections can be calculated are published annually by the European Association of Exploration Geophysicists, and computer programs are also widely available. Meter readings must be converted to gravity units before corrections are applied.

### 2.4.5 Drift corrections

After correction for tidal effects, *drift* can be calculated, usually by assuming that it has been linear in the time between two base readings. The first stage is to subtract the tidally-corrected initial reading at the drift base from all readings in the loop. This will give a first base reading of zero and a final one equal to the total drift. The sign of the correction is that needed to make this final reading zero also.

Drift corrections can be estimated graphically or be calculated, to a final

accuracy of 0.1 g.u. Absolute 'observed gravities' may be obtained by adding the absolute value at the base station to the drift-corrected gravity differences between the base and field stations.

### 2.4.6 Field notebooks

At each station, the number, time and meter reading must be recorded. Additional columns may be reserved for tidal and drift corrections, since drift should be calculated each day. Increasingly, however, calculations are made on microcomputers or programmable calculators and not in field notebooks.

Any factors which might affect a reading, such as heavy vibration from machinery, traffic, livestock or people, unstable ground or the possible presence of underground cavities, should be noted in the 'Remarks' column. Comments on weather conditions may also be useful, even if only as indicating the

34

Gravity field sheet (Fig. 2.11)

| OBSERVER | DATE | AREA | JOB No. |
|---|---|---|---|
| S.A.R.S.O.N.O | 07.08.86 | SI MEULUE | 8.6.1.1.5 |

| Base Station | Base Gravity | Meter | Scale Factor |
|---|---|---|---|
| S.I.N.A.B.A.N.G | 9.7.8.1.1.2 6.0.1 | L.G.7.5.9 | 1 |

| Station | Time | Reading | Tide | T.C. | Notes |
|---|---|---|---|---|---|
| 8.6.1.1.5.9.0.0.2 | 0.8.1.5 | 1.8.2.9.0.2 | | | Lhsseu river 14 m above river l. |
| 8.6.1.5.0.0.8.0.0 | 0.9.3.5 | 1.8.3.7.2.5 | | | " 1.5 m a.r.l. |
| 8.6.1.5.0.0.9.1 | 1.0.1.1 | 1.8.3.1.2.5.9 | | .0.5 | At river level, 1 m d 7 m |
| 8.6.1.5.0.0.9.2 | 1.0.3.1 | 1.8.4.1.4.8 | | | from banks 4 m high |
| 8.6.1.5.9.0.0.3 | 1.2.5.8 | 1.8.2.8.2.4 | | | Base check |
| 8.6.1.5.0.0.9.3 | 1.4.2.1 | 1.8.3.3.9.6 | | | 1.1 m a.s.l. Bangkaru Bay |
| 8.6.1.5.0.0.8.4 | 1.5.1.6 | 1.8.3.1.2.0.4 | | | Gabbro? outcrop at beach 20 cm a.s.l. |
| 8.6.1.5.0.0.8.5 | | | | | Sea level, mangrove (unstable ground) |
| 8.6.1.5.0.0.9.5 | 1.5.5.2 | 1.8.2.1.9.8 | | | 50cm a.s.l. at side of mosque |
| 8.6.1.5.0.0.9.6 | 1.6.1.0 | 1.8.2.8.7.7 | | | |
| 8.6.1.5.9.0.1.7 | 1.6.3.8 | 1.8.2.8.5.1 | | | New Sinabang wharf base |
| 8.6.1.5.9.0.0.2 | 1.6.4.3 | 1.8.2.9.0.1 | | | ABAB tie |
| 8.6.1.5.9.0.1.7 | 1.6.4.7 | 1.8.2.8.5.0 | | | to hotel base |
| 8.6.1.5.9.0.1.2 | 1.6.5.2 | 1.8.2.9.0.0 | | | |

**Fig. 2.11** Gravity field sheet designed for computer entry of data. Station numbering follows a widely used sequential system in which the first pair of digits show the year and the second pair identify the survey in that year. A '9' in the fifth column indicates a base station.

35

observer's state of mind. Where local terrain corrections are only occasionally significant, estimates may also be entered as 'Remarks'. Individual terrain-correction sheets may have to be prepared for each station in rugged areas.

Each loop should be headed with the observer's name or initials, the gravity-meter serial number and calibration factor and the base station number and gravity value. If the data are to be processed by computer, field sheets may be printed as coding-sheets for ease of data entry (Fig. 2.11); on these it may also be appropriate to record latitude and longitude, to the nearest degree, and the GMT–local time difference, for computer calculation of tidal effects.

Gravity data are expensive to acquire and deserve to be treated with respect. The general rules of Section 1.3.2 should be scrupulously observed.

### 2.4.7 *Elevation control*

Elevations of gravity survey points may be determined in many different ways. If 1-g.u. contours are required, high-accuracy optical or radio-wave techniques are essential, while for the 50 or 100 g.u. contours common in regional surveys, barometric levelling or direct reference to sea-level and tide tables may be adequate. Obtaining elevation control is likely to be the most expensive part of a gravity survey, and advantage should always be taken of any 'free'

levelling done for other purposes (e.g. in pegging seismic lines).

### 2.5 Field interpretation

Gravity data are now usually interpreted by calculating the fields produced by geological models and comparing these with the actual data. This requires some form of computer and is not generally done in the field. However, an appreciation of the effects associated with a few simple bodies can help the field observer to assess the validity and significance of the data being collected. Such an understanding can sometimes lead to a vital decision to infill with additional stations being taken at a time when this can be done quickly and economically.

### 2.5.1 *The Bouguer plate*

The *Bouguer plate* provides the simplest possible interpretational model. An easily memorized 'rule-of-thumb' is that the gravity field due to a slab of material 1-km thick with a density contrast of 100 $kg \cdot m^{-3}$ with its surroundings is about 40 g.u. The effect varies in direct proportion to both thickness and density contrast. The relationship is valid even for readings taken some distance above the upper surface, provided that the distance of the station from the nearest edge of the slab is large compared with its height above the lower surface.

## Example 2.1

If the standard crustal density is taken to be 2670 kg·m$^{-3}$, the effect of the upper sediment layer, 1.5 km thick, in Fig. 2.12 would be $1.5 \times 3.7 \times 40 = 220$ g.u. (approximately) at the centre of the basin, and of the deeper sediments would be $1.6 \times 2.7 \times 40 = 170$ g.u. (approximately). The total (negative) anomaly would be about 390 g.u.

### 2.5.2 Spheres and cylinders

Less-extensive bodies can be modelled by homogeneous spheres or by homogeneous cylinders with circular cross-sections and horizontal axes. The field due to a sphere, radius $r$ measured at a point immediately above its centre, is:

$$g = \frac{4}{3}\rho\pi G\,\frac{r^3}{h^2}$$

The factor $\frac{4}{3}\rho\pi G$ is about 28 g.u. for a density contrast of 100 kg·m$^{-3}$ and lengths measured in kilometres. The depth, $h$, to the centre of the sphere is roughly equal to four-thirds of the half-width of the anomaly.

For a horizontal cylinder of circular cross-section, the corresponding field is:

$$g = 2\rho\pi G\,\frac{r^2}{h}$$

For 100 kg·m$^{-3}$ density contrast and distances in kilometres, $2\rho\pi G$ is about 40. The half-width of the anomaly is equal to the depth of the axis of the cylinder.

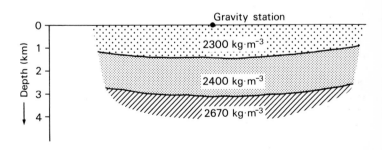

**Fig. 2.12** Sedimentary basin model, suitable for Bouguer-plate methods of approximate interpretation. Note that basement is assigned the standard crustal density of 2670 kg · m$^{-3}$.

## Example 2.2

Interpreting the anomaly of Fig. 2.13 as due to a roughly spherical, air-filled cavity in rock of density 2500 $kg \cdot m^{-3}$:

Amplitude of anomaly $\simeq$ 0.45 g.u. Half-width $\simeq$ 2 m (i.e. $2.0 \times 10^{-3}$ km). Therefore depth to sphere centre = $2.7 \times 10^{-3}$ km

$$r^3 = \frac{\text{gravity anomaly} \times h^2}{28 \times (\text{density contrast}/100)}$$

$$= \frac{0.45 \times 2.7^2 \times 10^{-6}}{28 \times 25}$$

$$r = 1.7 \times 10^{-3} \text{ km (i.e. 1.7 m)}$$

## Example 2.3

Interpreting the anomaly in Fig. 2.13 as due to a roughly cylindrical air-filled cavity in rock of density 2500 $kg \cdot m^{-3}$:

Amplitude of anomaly $\simeq$ 0.45 g.u. Half-width $\simeq$ 2 m (i.e. $2.0 \times 10^{-3}$ km) Therefore depth to cylinder axis = $2.0 \times 10^{-3}$ km

$$r^2 = \frac{\text{gravity anomaly} \times h}{40 \times (\text{density contrast}/100)}$$

$$= \frac{0.45 \times 0.002}{40 \times 25}$$

$$r \simeq 0.001 \text{ km (i.e. 1 m)}$$

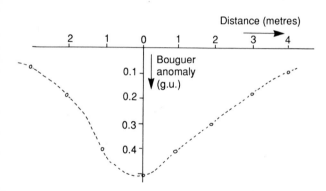

**Fig. 2.13** Detailed Bouguer anomaly profile over a sub-surface cavity.

### 2.5.3 Nettleton's method for direct determination of density

The *bulk density of topography* may be determined by a method first suggested by Nettleton. This assumes the correct density to be the one which removes the effect of topography from the gravity map when corrections are made. The method can only work if there is no real gravity anomaly associated with the

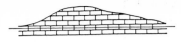

SUCCESS

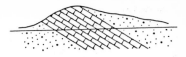

FAILURE

**Fig. 2.14** Example of cases in which Nettleton's method of determining density could be expected to succeed and fail.

topography and will fail if, for example, a hill is the surface expression of a dense igneous plug or a dipping limestone bed (Fig. 2.14). Nettleton's method may be applied to a profile or to all the gravity stations in an area. In the latter case, a computer may be used to determine the density value which produces the least correlation between topography and the corrected anomaly map.

Field observers should be aware of the Nettleton technique since opportunities may arise to take additional readings for density control.

# 3

# Magnetic method

Compasses and dip-needles were used to find magnetite in Sweden in the Middle Ages, making the magnetic method the oldest of all applied geophysical techniques. It is still one of the most widely used, even though significant magnetic effects are produced by only a very small number of minerals.

## 3.1 Magnetic properties

Although governed by the same fundamental equations, magnetic and gravity surveys are very different. The magnetic properties of rocks may differ by several orders of magnitude, rather than by merely a few tens of percent.

### 3.1.1 Poles, dipoles and magnetization

The isolated magnetic pole, which would produce a field obeying the inverse-square law, is merely a convenient fiction. The fundamental magnetic source is the dipole, but since a line of consecutive dipoles produces the same effect as would positive and negative poles isolated at opposite ends of the line (Fig. 3.1), the pole concept is often useful.

A dipole placed in a magnetic field tends to rotate, and so is said to have a *magnetic moment*. The moment of the simple magnet of Fig. 3.1, which is effectively a positive pole, strength *m*, at a distance $2L$ from a negative pole $-m$, is equal to $2Lm$. The magnetization of a solid body is defined in terms of the magnetic moment per unit volume, and is a vector, having direction as well as magnitude.

### 3.1.2 Susceptibility

A body placed in a magnetic field acquires a magnetization which, if small, is proportional to the field,

$$M = kH$$

The susceptibility $k$ is very small for most natural materials, and may be either negative (diamagnetism) or positive (paramagnetism). Only in exceptional circumstances do the fields produced by such materials

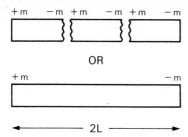

**Fig. 3.1** Combination of magnetic dipoles to form a bar magnet with magnetic moment $2\,Lm$.

affect survey magnetometers.

Observable magnetic anomalies are due to a small number of substances which are ferro- or ferri-magnetic. In these, the molecular magnets are held parallel by intermolecular 'exchange forces' which, below the Curie temperature, are stronger than the effects of thermal agitation. Magnetite, pyrrhotite and maghemite, all of which have Curie temperatures of about 600°C, are the only three important naturally-occurring magnetic minerals and, of the three, magnetite is by far the most common. Hematite, the most abundant iron mineral, has a very small susceptibility, and therefore many iron ore deposits do not produce magnetic anomalies.

The magnetic properties of very magnetic rocks tend to be extremely variable, and their magnetization is not strictly proportional to the applied field. Quoted susceptibilities are always for Earth-average field strengths.

### 3.1.3 Remanence

Ferro- and ferri-magnetic materials may have permanent as well as induced magnetic moments, so that their magnetization is not necessarily in the direction of the Earth's field.

The magnitudes of induced and permanent moments can be compared using the *Konigsberger ratio* of the permanent moment to the moment that would be induced in an Earth-standard field of 50 000 nT (nanoTesla). The ratio is generally large in highly magnetic rocks and small in weakly magnetic ones, but hematite occasionally has an extraordinarily high ( > 10 000) Konigsberger ratio and magnetic anomalies due entirely to remanence are sometimes produced by hematitic ores.

### 3.1.4 Susceptibilities of rocks and minerals

The *susceptibility* of a rock usually depends only on its magnetite content. Sediments and acid igneous rocks have small susceptibilities whereas basalts, dolerites, gabbros and serpentinites are usually strongly magnetic. Weathering generally reduces susceptibility because of the oxidation of magnetite to hematite, but some laterites are magnetic because of the presence of maghemite and remanently-magnetized hematite.

Susceptibility ranges of some common rocks and minerals, in rationalized SI units, are given in Table 3.1.

**Table 3.1** Magnetic susceptibilities of common rocks and ores (e.m.u.)

*Common rocks*

| | | | |
|---|---|---|---|
| Slate | 0–0.002 | Dolerite | 0.01–0.15 |
| Greenstone | 0.0005–0.001 | Basalt | 0.001–0.1 |
| Granulite | 0.0001–0.05 | Rhyolite | 0.00025–0.01 |
| Salt | 0.000–0.001 | Gabbro | 0.001–0.1 |
| Limestone | 0.00001–0.0001 | | |

*Ores*

| | | | |
|---|---|---|---|
| Hematite | 0.001–0.0001 | Magnetite | 0.1–20.0 |
| Chromite | 0.0075–1.5 | Pyrrhotite | 0.001–1.0 |
| Pyrite | 0.00010.005 | | |

## 3.2 The magnetic field of the Earth

Earth magnetic field strengths are now usually quoted in nanoTesla (nT), numerically equal to an older unit, the gamma (0.00001 gauss), which is still often used on maps and in conversation.

The magnetic fields of geological bodies are superimposed on the background of the Earth's main field. Variations in magnitude and direction of this field influence both the magnitudes and shapes of local anomalies.

The 'North' and 'South' poles of common usage are replaced in geophysics by positive and negative. Magnetic field direction is conventionally defined as the direction in which a unit positive pole would move but, since all things are relative, geophysicists give little thought to whether it is the North or South magnetic pole which is positive.

### 3.2.1 Characteristics of the Earth's main field

The Earth's main field, which originates in electric currents in the liquid outer core, is approximately dipolar but changes slowly in both magnitude and direction. The dipole field is described in Section 1.1.5 and its characteristics are plotted in Fig. 3.2. A magnetic dipole at the Earth's centre, inclined at about 11° to the spin axis, would account for most of the field observed at the Earth's surface. Distortions extending over areas several hundreds to thousands of kilometres across can be thought of as caused by a relatively small number of subsidiary dipoles at the core–mantle boundary.

Neither the magnetic equator, which links points on the Earth's surface of zero magnetic dip, nor the magnetic poles coincide with their geographic equivalents (Fig. 3.3). Differences between the directions of true and magnetic North are termed

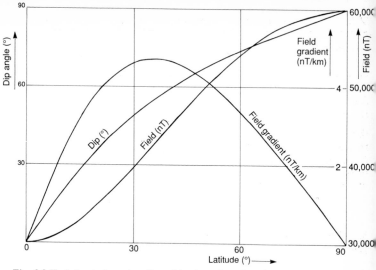

**Fig. 3.2** Variation in intensity, dip and horizontal gradient for an ideal Earth-dipole aligned along the Earth's spin axis and producing a polar field of 60 000 nT.

declinations, presumably because a compass needle *ought* to point North but *declines* to do so. Because of their navigational importance, declinations are usually to be found listed in the keys which accompany conventional topographic maps.

Dip angles near the equator change twice as fast as latitude angles. Dips estimated from Fig. 3.3 can be used, in conjunction with Fig. 3.2, to obtain rough estimates of regional gradient. Such estimates are useful in determining whether a gradient is likely to be significant but give only approximate correction factors. Corrections have East–West as well as North–South components, the gradient being roughly parallel to the local Magnetic North arrow, but since they generally amount to

only a few nanoTeslas per kilometre, they are usually ignored in ground work.

### 3.2.2 The International Geomagnetic Reference Fields (IGRF)

The variations of the Earth's main field with latitude, longitude and time are described by complicated, experimentally-determined equations which define the International Geomagnetic Reference Fields (IGRF). These are quite good representations of the actual fields in well-surveyed areas, where they can be used to calculate regional corrections, but there are discrepancies of as much as 250 nT in areas from which little information was avail-

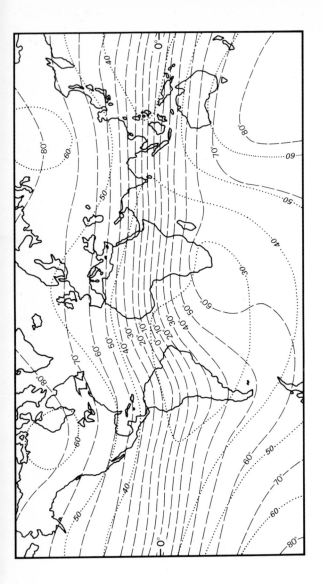

**Fig. 3.3** Dip and intensity of the Earth's magnetic field (IGRF). Dashed lines show dip, in degrees; dotted lines show intensity, in thousands of nanoTeslas (nT).

able at the times of formulation. Local reference fields are used as alternatives in some countries, Australia being a notable example.

Because long-term changes ('secular' changes) are not predictable except by extrapolation from past observations, the IGRF must be updated every decade and is also revised retrospectively. Fortunately, calculated time-dependent corrections, important when comparing airborne or marine surveys carried out months or years apart, are less vital in ground work where individual stations can be re-occupied.

### 3.2.3 Diurnal variations

The Earth's magnetic field also varies because of changes in the strength and direction of currents circulating in the ionosphere. In the normal pattern, termed solar-quiet (Sq), the background field is almost constant during the night but decreases between dawn and about 11 a.m., increases again until about 4 p.m. and then slowly declines to the overnight value (Fig. 3.4). The peak-to-trough amplitudes in mid-latitudes are of the order of a few tens of nanoTeslas. The variations, being dependent on solar radiation, tend to be directly related to local solar time but, for points up to a few hundred kilometres apart, amplitude differences due to differences in crustal conductivity may amount to more than 20% and are often more important than time dependency.

Within about 5° of the Magnetic Equator the diurnal variation is strongly influenced by the *equatorial electrojet*, a band of high conductivity in the ionosphere about 600-km wide. Near the Magnetic Equator the daily variations may be well in excess of 100 nT and there may be differences of 10 to 20 nT in the amplitudes of the diurnal curves at points only a few tens of kilometres apart.

Most of the magnetic phenomena

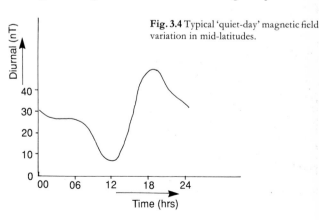

**Fig. 3.4** Typical 'quiet-day' magnetic field variation in mid-latitudes.

in polar regions can be explained in terms of an *auroral electrojet* which is subject to severe short-period fluctuations. In such regions it is particularly important that background variations be monitored directly. Returning to a base station at intervals of one or two hours may be quite insufficient.

### 3.2.4 *Magnetic storms*

Short-term auroral effects are special cases of the irregular disturbances (Ds and Dst) known as *magnetic storms*. These are produced by sunspot and solar flare activity and, despite their name, are not meteorological, often occurring on clear, cloudless days. There is usually a sudden onset, during which the field may change by hundreds of nano-Teslas, followed by a slower, erratic return to normality. Time scales vary widely, but the effects persist for periods of hours and often days.

Ionospheric prediction services in many countries give advance warnings of the general probability of storms but not of their detailed patterns, and the field changes in both time and space are too rapid for corrections to be applied. Survey work has to stop until a storm is over.

Aeromagnetic data are severely affected by quite small irregularities and for contract purposes *technical magnetic storms* are defined, sometimes as departures from linearity in the diurnal curve of as little as 5 nT in an hour.

### 3.2.5 *Geological effects*

The Curie point for all geologically important magnetic materials is at about 550–600°C. This temperature is reached in the lower part of normal continental crust but within the mantle beneath the oceans. The upper mantle is only weakly magnetic, so that the effective base of local magnetic sources is the Curie isotherm in continental areas and the Moho beneath the oceans.

Magnetic fields due to massive magnetite deposits can amount to as much as 200 000 nT, several times the magnitude of the Earth's normal field. Because of the dipolar nature of magnetic sources, these, and all other, magnetic anomalies have positive and negative parts, so that in some cases directional magnetometers record negative fields. Anomalies of this size are very unusual, but basalt dykes and flows and some larger basic intrusions produce fields of thousands and occasionally tens of thousands of nanoTeslas. Anomalous fields of more than 1000 nT are otherwise rare, even in areas of outcropping igneous basement, and sedimentary rocks generally produce changes of less than 10 nT.

In some tropical areas, magnetic fields of tens of nanoTeslas are produced by maghemite formed as nodular growths in laterites. The nodules may later weather out to form ironstone gravels which have marked effects on hand-held magnetometers. The factors which control the formation of maghemite rather than the commoner, non-magnetic form of haematite are as yet not fully understood. Fields from such sources are usually regarded as 'noise' (see Section 3.4.6).

## 3.3 Magnetic instruments

Early magnetometers used compass needles mounted on horizontal axes (dip needles) to measure vertical field. These 'torsion' magnetometers were in common use until about 1960, when they began to be replaced by ground versions of the instruments used for airborne survey, the fluxgates and proton precession magnetometers.

### 3.3.1 The fluxgate

The sensing element of a fluxgate magnetometer consists of one or more cores of magnetic alloy, around which are wound coils through which alternating current can be passed. Variations in the electrical properties of the circuits with magnetization of the cores can be converted into voltages proportional to the external magnetic field along the core axes. Measurements are thus of the magnetic field component in whichever direction the sensor is pointed. In ground instruments a bull's-eye levelling bubble is usually set in the casing and vertical fields are measured.

Fluxgates do not measure absolute fields and so require calibration. They are also subject to thermal drift since the magnetic properties of the cores and the electrical properties of the circuits all vary with temperature. Ground instruments, for which portability is of primary importance, have only limited thermal insulation and may be accurate to only 10 or 20 nT. In recognition of this, readings are often displayed, rather crudely, by the position of a needle on a graduated dial. Despite claims to the contrary by some manufacturers, such sensitivity is quite inadequate for much ground survey work.

One problem that was not anticipated when many of the portable fluxgates were designed was that non-magnetic batteries would become almost unobtainable. Most dry batteries are now steel-jacketed and make nonsense of the readings if installed in the same housing as the sensor. External battery packs can be used, but these are clumsy and reduce rather than eliminate the effects. The basic problem is that, because the observer must level the instrument, he, and anything magnetic he may be carrying, will be close to the sensor when the reading is taken.

### 3.3.2 Proton precession magnetometer

The proton magnetometer makes use of the small magnetic moment of the hydrogen nucleus (proton). The sensing element consists of a bottle containing a low freezing-point hydrocarbon fluid, about which is wound a coil of copper wire. A current of the order of an Ampere or more is passed through the coil, creating a strong magnetic field along which the proton moments are aligned. Although many fluids *can* be used, the manufacturer's recommendation, usually for high-purity decane, should always be followed if the bottle has to be 'topped-up'.

When the current is switched off, the protons tend to re-align in the direction of the external field. In classical mechanical terms, they precess about the field direction, as a gyroscope precesses about the Earth's gravity field. The precession frequency is proportional to the field strength.

In Quantum Theory, the re-orientation occurs as an abrupt 'flip' with the emission of a quantum of electromagnetic energy. Both theories relate the frequency of the electromagnetic waves to the external field via two of the most accurately known of all physical quantities, Planck's constant and the proton magnetic moment. In the Earth's field of about 50 000 nT, the precession frequency is about 2000 Hz. Sophisticated phase-sensitive circuitry is needed to measure such frequencies to accuracies of one part in fifty thousand in the half-to-one second which is all that modern geophysicists will tolerate.

In theory the proton magnetometer is capable of almost any desired accuracy, but in practice the need for short reading times and reasonable polarization currents sets the limit at 1.0 or 0.1 nT. Ultra-portable versions are available which read to only 10 nT sensitivity but these are only marginally easier to use and have not become popular.

The proton magnetometer measures only total fields, which can cause problems in interpreting large anomalies in which the direction of the resultant field changes rapidly from place to place. Also, erratic readings may be caused by interference from power lines and even from eddy currents induced in nearby conductors by the termination of the polarizing current, and also by strong field gradients. However, these are only minor drawbacks and the 1 nT or 0.1 nT proton magnetometer is now by far the commonest instrument in ground surveys, as well as in the air and at sea. The self-orientation property allows the sensor to be supported on a staff well away from both the observer and from small magnetic sources at ground level (see Fig. 1.5) and it is also an advantage that readings are obtained as drift-free absolute values directly in nano-Teslas, even though corrections must still be made for diurnal variations.

### 3.3.3 Memory magnetometers

Proton magnetometers are now available in which data can be stored for later recall to the front-panel display or for transfer directly to a printer or microcomputer. Each stored reading (of which there may be more than a thousand) has a serial number and may also be associated with a station number which can be automatically incremented by an amount equal to the station spacing. The times at which readings are taken are also recorded. Readings can be initiated by the clock circuitry at regular intervals, usually of between 5 and 999 seconds, allowing diurnal variation to be monitored. The need for bulky and unreliable pen recorders at unmanned base stations has thus now been eliminated.

With memory instruments there is no absolute need to record readings in field notebooks. However, if books are not used for the magnetic data, observers are much less likely to make notes describing the magnetic, geological or topographic environment. A limited number of descriptive codes can be entered into the memories of some instruments, but it is hardly practicable to cover the full range of possible field situations in this way.

### 3.4 Magnetic survey practice

Although absolute numerical readings are obtained (and can be repeated) at the touch of a button in a proton magnetometer survey, faulty magnetic maps can be produced if simple precautions are ignored.

All base locations, whether used for repeat readings or for continuous diurnal monitoring, should be checked for field gradients. If moving the sensor a metre produces a significant change in reading, another base point should be sought.

*3.4.1 Beginning a survey*

The first stage in any magnetic survey is to check the magnetometers (and the operators). Operators can be potent sources of magnetic noise, although the problems are much less acute with proton magnetometer sensors on 3-m poles than with fluxgate instruments which are held close to the body. With a magnetometer used as in Fig. 1.5, it is quite unlikely that there will be effects from anything carried by the operator, but serious errors can occur using the alternative 'hands-off' technique with the sensor in a back-pack. Compasses, pocket knives and geological hammers are all detectable if within about a metre of the sensor. Survey vehicles can affect results at distances of up to 20 m and their magnetic properties should be determined before starting survey work.

As absolute instruments, all proton magnetometers should give the same reading at the same time in the same place. Differences were often greater than 10 nT between instruments manufactured prior to 1980 but are now seldom more than 1 or 2 nT. Sensor 'bottles' can be placed very close together and may even touch when checks are being made, but readings cannot be completely simultaneous as there would then be interference between the two polarizing fields.

Large discrepancies and very variable readings usually indicate that a magnetometer is poorly tuned. The correct tuning range can be roughly identified using maps of the type shown in Fig. 3.3, but final checks should always be made in the field.

If the tuning is roughly correct, near-identical readings should be obtained if the setting is varied over a range of about 10 000 nT. The correct setting is given by the actual values observed (e.g. 47 000 in Example 3.1). Manual magnetometers are generally rather coarsely tunable in steps of a few thousand nanoTeslas, but greater accuracy is possible with micro-processor control. It is partly this finer control which allows memory-magnetometers to be routinely read to 0.1 nT, and a warning of faulty tuning or high gradient is given by failure to display a digit beyond the decimal point.

It can be seen that repeatability alone is no guarantee of correct tuning. It is the range of settings over which the circuits can 'lock-on' to the pre-cession signal which provides the crucial evidence.

### 3.4.2 Monitoring diurnal variation

If only a single instrument is available, diurnal corrections must rely on repeated visits to a base or sub-base, ideally at intervals of less than one hour.

A more complete diurnal curve can be constructed using an extra magnetometer to obtain readings at a fixed point at 3 to 5 minute inter-vals. In principle it is then possible to dispense with base station re-occu-pations by field magnetometers, except for one or two comparisons with the diurnal monitor. It is, however, poor practice to rely

**Example 3.1: Magnetometer tuning (manual model)**

| Tuning setting | Readings | | |
|---|---|---|---|
| 30 000 | 31 077 | 31 013 | 31 118 |
| 32 000 | 32 770 | 32 788 | 32 775 |
| 34 000 | 35 055 | 34 762 | 34 844 |
| 36 000 | 37 481 | 37 786 | 37 305 |
| 38 000 | 42 952 | 40 973 | 41 810 |
| 41 000 | 47 151 | 47 158 | 47 159 |
| 44 000 | 47 160 | 47 158 | 47 156 |
| 47 000 | 47 171 | 47 169 | 47 169 |
| 50 000 | 47 168 | 47 175 | 47 173 |
| 53 000 | 47 169 | 47 169 | 47 169 |
| 56 000 | 53 552 | 54 602 | 54 432 |
| 60 000 | 59 036 | 59 292 | 58 886 |
| 64 000 | 65 517 | 65 517 | 65 517 |

entirely on the monitor, since recov-ery of field data after a failure will then be difficult, if not impossible. Problems are all too common even when the monitor is manually oper-ated, but especially likely with unattended 'memory' instruments.

The battery drain of an automatic magnetometer is rather high and in some models the transition from operational to unworkable occurs suddenly and without warning. Readings already stored are pre-served by the action of a separate 'keep-alive' lithium battery but diurnal control will be lacking for the rest of the day.

Especial vigilance is needed if field and base instruments can be linked by a data-exchange line at the end of a period of work and corrections are made automatically. Absurdities in

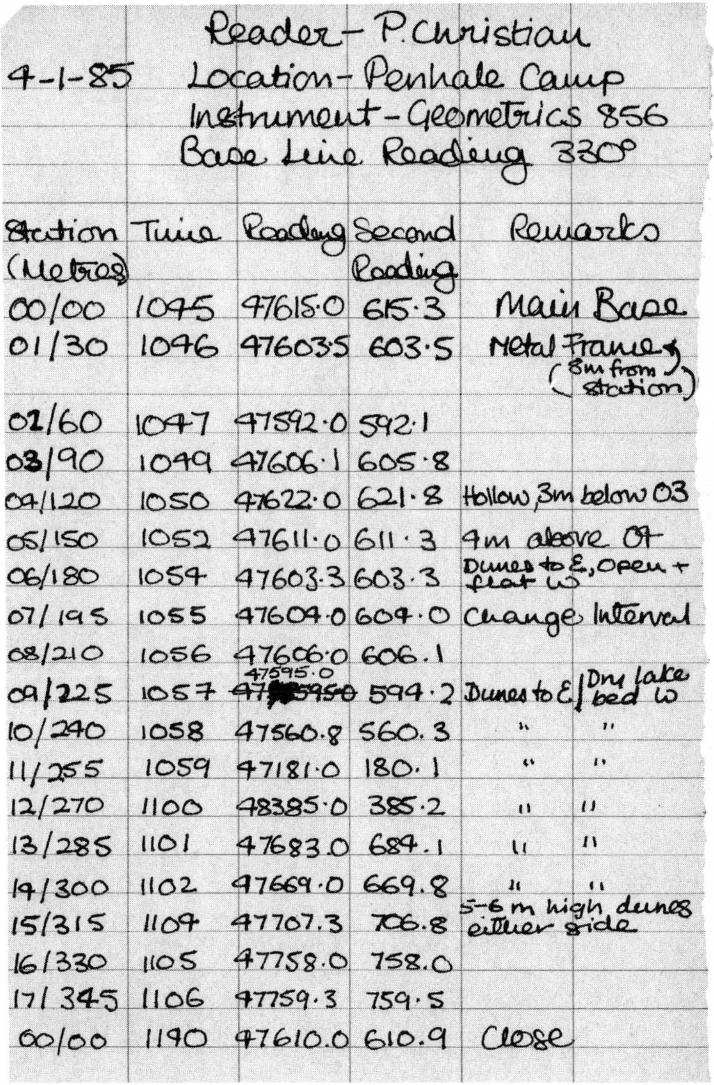

Reader - P. Christian

4-1-85   Location - Penhale Camp
          Instrument - Geometrics 856
          Base Line Reading 330°

| Station (Metres) | Time | Reading | Second Reading | Remarks |
|---|---|---|---|---|
| 00/00 | 1045 | 47615·0 | 615·3 | Main Base |
| 01/30 | 1046 | 47603·5 | 603·5 | Metal Frames (8m from station) |
| 02/60 | 1047 | 47592·0 | 592·1 | |
| 03/90 | 1049 | 47606·1 | 605·8 | |
| 04/120 | 1050 | 47622·0 | 621·8 | Hollow, 3m below 03 |
| 05/150 | 1052 | 47611·0 | 611·3 | 4m above 04 |
| 06/180 | 1054 | 47603·3 | 603·3 | Dunes to E, open + flat W |
| 07/195 | 1055 | 47604·0 | 604·0 | Change Interval |
| 08/210 | 1056 | 47606·0 | 606·1 | |
| 09/225 | 1057 | ~~47595·0~~ 47595·0 | 594·2 | Dunes to E / Dry lake bed W |
| 10/240 | 1058 | 47560·8 | 560·3 | " " |
| 11/255 | 1059 | 47181·0 | 180·1 | " " |
| 12/270 | 1100 | 48385·0 | 385·2 | " " |
| 13/285 | 1101 | 47683·0 | 684·1 | " " |
| 14/300 | 1102 | 47669·0 | 669·8 | " " |
| 15/315 | 1104 | 47707·3 | 706·8 | 5-6 m high dunes either side |
| 16/330 | 1105 | 47758·0 | 758·0 | |
| 17/345 | 1106 | 47759·3 | 759·5 | |
| 00/00 | 1190 | 47610·0 | 610·9 | Close |

**Fig. 3.5** Magnetometer field notebook; a typical page.

51

the diurnal data (such as might be caused by an inquisitive passer-by driving up to the base and examining the magnetometer) may then not be detected until a late stage in map production, if at all.

Obviously, bases should be remote from possible sources of magnetic interference (especially temporary sources such as traffic), and their locations should be describable for future reference.

### 3.4.3 Standard values

Diurnal curves show how field strength has varied at one fixed point, and data processing is simplified if the monitor is set up at the same point throughout the survey. A *standard value* needs to be allocated to this point, preferably by the end of the first day of survey work. The choice is to some extent arbitrary; if the variation in measured values were to be between 32 340 nT and 32 410 nT it might be convenient to adopt 32 400 nT as standard, even though this was neither the mean nor the commonest value. Alternatively, the standard diurnal value might be chosen to correspond to a convenient value at the main reference base.

Unless a survey area is so small that a single reference base can be used, a number of sub-bases will have to be established and their standard values determined (see Sections 1.3.3 and 1.3.4). The underlying principle is that if, at a time when the base magnetometer actually records the diurnal standard value, identical instruments are read at all other bases and sub-bases, these will all record the appropriate standard values. The field data can then be processed so that this will be true of the corrected values assigned to all survey points.

### 3.4.4 Field survey procedure

At the start of each survey day the diurnal magnetometer should be set up in the chosen location, which ideally should be one that vehicles cannot possibly approach. The first reading of each day should be at a base or sub-base, and should be made at the same time as a reading is being taken, either automatically or by hand, at the diurnal magnetometer. This does not necessarily imply that the two instruments must be adjacent at that time.

All readings, whether of bases or singly-occupied field points, should be quickly repeated and the two readings should differ by not more than 1 nT. Greater differences may indicate high field gradients, which may need to be investigated further. Large differences between readings at adjacent stations call for 'fill-in' at intermediate points; it is obviously desirable that the operator notices this and infills immediately.

A page from a typical field notebook is shown in Fig. 3.5. At each station the location, time and reading must be recorded, as well as any relevant topographic or geological information and details of any visible or suspected magnetic sources. Unless the grid is already well

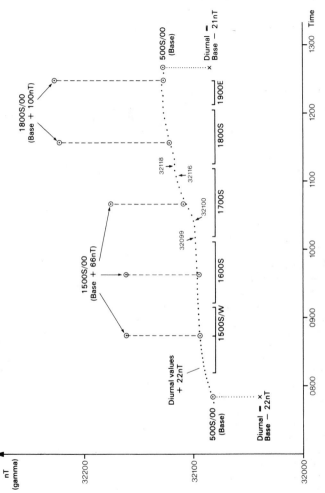

**Fig. 3.6** Diurnal corrections. The diurnal has been both recorded by a memory instrument and monitored by repeat readings at various base stations. Time periods during which various field traverses were being observed are also shown.

mapped, the notebook should contain enough information for the lines to be positioned on maps or airphotos.

At the end of the day, a final reading should be made at the first base to be occupied. This should again be timed to coincide with a reading of the diurnal magnetometer. It is good practice to transcribe the diurnal data in to the field notebook, which then contains a complete record of the day's work.

### 3.4.5 Processing magnetic data

During the survey, bases or sub-bases should be occupied at intervals of not more than two hours, so that data can be processed even if the diurnal record is lost or proves faulty. The way in which such readings might be used to provide diurnal control is shown in Fig. 3.6, which also shows the automatically recorded diurnal curve. The standard value for the main base in this area differed from that at the diurnal magnetometer by 22 nT, as was confirmed by the readings at the beginning and end of the day. It can be seen that the greatest error in using straight-line interpolation between diurnal values derived from field bases would be about 5 nT and would affect Line 1700S. Interpolation using a smooth curve instead of straight lines would significantly reduce this error.

Using standard values, magnetic data can be corrected for diurnal variation in two different ways. The diurnal correction at any time is simply the difference between the diurnal standard value and the actual diurnal reading. The most straight-forward approach is to determine, by interpolation when necessary, the diurnal value at the time a given field reading was made. The diurnal standard value can be subtracted from this and the result can then be subtracted from the field reading. If a microcomputer is available at the field base, the whole operation can be automated.

Although this method is simple in principle and provides individual values at all field points, it is tedious and error-prone if a hundred or more stations have to be processed by hand each evening. If instead contouring is based on 'cuts' on profiles of uncorrected readings, fewer calculations are needed and errors and peculiarities in the data are immediately obvious.

Figure 3.7 shows the uncorrected profile for Line 1700S of Fig. 3.6, together with a base diurnal line obtained by adding the 22 nT difference between main reference base and diurnal monitor point to diurnal values recorded automatically at 5-minute intervals. Since the reference-base standard value is 32 100 nT, the points at which this diurnal line intersects the profile correspond to points on the ground where the magnetic field also has a corrected value of 32 100 nT.

A second, parallel, diurnal line identifies points at which the field is 50 nT higher than at the base, and so on. Only these 'contour cuts' need

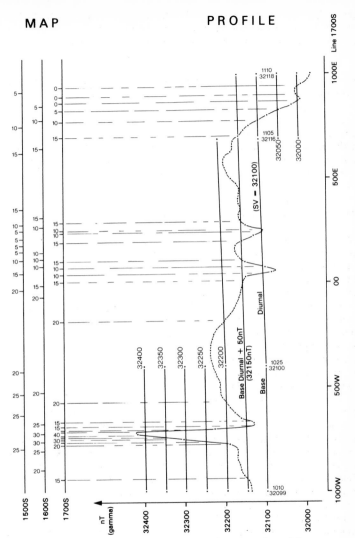

**Fig. 3.7** PROFILE shows contour cuts by diurnal curve and parallel curves on a profile of uncorrected magnetic data. MAP shows contour cuts plotted on traverse lines for contouring. Transfer process illustrated for line 1700S, with cuts recorded in tens of nanoTeslas, i.e. a cut of 10 corresponds to a value of 32 100 nT.

be plotted on the map, and where they are very close together some may be omitted. Contouring becomes very simple.

It is clear that, with contours at 50-nT intervals, the 5-nT discrepancy between the diurnal curves based on direct observation and on base reoccupations is unimportant. Profile displays are excellent for assessing the significance, or insignificance, of diurnal effects and noise, and their preparation should be a field priority. This is true even if microcomputers are used to compute individual values at each field point, but curves can then be drawn of corrected rather than raw data.

### 3.4.6 Noise in ground magnetic surveys

Magnetic readings in populated areas are usually affected by stray fields from pieces of iron and steel. Even if no such materials are visible, profiles obtained along roads are usually very distorted compared to those obtained on parallel traverses through open fields only 10 or 20 m away. Since the sources are often quite small and lie within a metre or so of the ground surface, the effects produced are highly variable.

One approach to the noise problem is to try to take all readings well away from obvious sources, noting in the field books where this has not been possible. Alternatively, the almost universal presence of ferrous 'noise' can be accepted and the data can be filtered. For this approach to be successful, many more readings must be taken than would be needed to define the purely

geological anomalies. The technique is becoming more popular with increasing use of memory instruments which discourage note-taking but with which the additional field effort involved in taking four or five times the normal number of readings is minimal. It is only safe to use this method and dispense with notebooks if the survey grid is already well surveyed, well described and accurately located.

## 3.5 Simple magnetic interpretation

Field interpretation of magnetic data enables areas needing infill or of suspect data to be identified and then checked without the expense of a second survey at some later date. Good interpretation requires profiles, which preserve all the detail of the original readings, and contour maps which allow trends and patterns to be identified.

### 3.5.1 Forms of magnetic anomaly

The shape of a magnetic anomaly varies dramatically with the dip of the Earth's field, as well as with variations in the shape of the source body and its direction of magnetization. Simple 'pole' models can be used to obtain quick visual estimates of the form of anomaly produced by any magnetized body.

Figure 3.8a shows an irregular mass magnetized by induction in a field dipping at about 60°. Since the field direction defines the direction in which a positive pole would move,

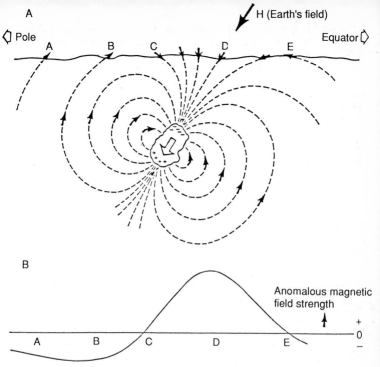

**Fig. 3.8** Production of a total field anomaly by induced magnetism. (*a*) The induced field; (*b*) anomaly profile.

the effect of the external field is to produce the distribution of poles shown diagrammatically. The secondary field due to these poles is indicated by the dashed lines of force. Direction is dictated by the simple rule that like poles repel.

If the secondary field is small, so that total-field readings are made in the direction of the external field vector, no anomalous field will be detected near C and E and the anomaly will be positive between these points and negative for con-

siderable distances beyond them. The anomaly maximum will be near D, giving a magnetic profile with its peak offset towards the Magnetic Equator (Fig. 3.8b).

At the Equator the total-field anomaly would be negative, centred over the body and would have positive side-lobes to North and South.

Because each positive magnetic pole is somewhere balanced by a negative, the net flux involved in any anomaly is zero. The fields from positive and negative poles will cancel over most parts of a uniform magnetized sheet and only the edges will be detected by a magnetic survey. Strongly magnetized bodies thus sometimes produce little or no anomaly.

### 3.5.2 'Rule of thumb' depth estimation

Depth estimation is one of the main objectives of magnetic interpretation. Very simple rules exist which can give depths to the tops of source bodies which are usually correct to within about 20% and are quite adequate for preliminary assessment of field results.

In Fig. 3.9 the part of the anomaly profile over which the variation is almost linear is emphasized by a thickened line. For many roughly

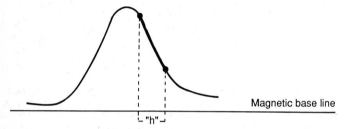

**Fig. 3.9** Straight-slope method of magnetic interpretation. '$h$', the distance over which the variation is linear, is often roughly equal to the depth to the top of the magnetized body.

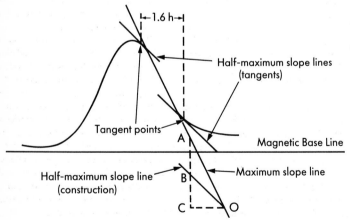

**Fig. 3.10** Peters' half-slope magnetic interpretation method. In the triangle AOC, AB = BC.

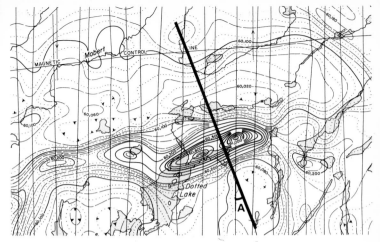

**Fig. 3.11** Effect of strike on magnetic interpretation. If the profile recorded along a flight line (one of the set of continuous, approximately straight lines) is interpreted directly, the depth estimate must be multiplied by the cosine of the angle A made with the (thick) line at right angles to the strike of the magnetic contours. (Example from aeromagnetic map of northern Canada.)

tabular and abruptly-truncated bodies, the depth to the top surface is roughly equal to the horizontal extent of the straight-line section. Although effective, this method is open to objection since there is actually no straight segment of the curve and the estimate relies on an optical illusion.

A slightly more complicated rule known as *Peters' method* has rather more of a theoretical basis. A tangent is drawn to the profile at the point of steepest slope, and lines with half this slope are drawn using the geometrical construction of Fig. 3.10. The two points at which the half-slope lines are tangents to the anomaly curve are found by eye or with a parallel ruler, and the hori-

zontal distance between them is measured. This distance is divided by 1.6 to give a rough depth to the top of the source body.

Peters' method relies on a range of model studies which showed that the true factor lies between about 1.2 and 2.0, with values close to 1.6 being common. In practice, the results will usually differ only slightly from those obtained using straight-slope methods. In either case the profile is assumed to have been measured along a line at right angles to the strike of the anomaly, and in all other cases the estimate must be multiplied by a factor equal to the cosine of the intersection angle (A in Fig. 3.11).

# 4

# Radioactivity surveys

The radioactivity of rocks is now chiefly monitored using gamma-ray scintillometers and spectrometers. Although most radiometric instrumentation was developed with uranium search in mind, secondary uses were soon found. Amongst these are regional geological mapping and correlation, exploration for some industrial minerals and *in situ* determinations of phosphates. The same instruments may also be used to track the movement of artificial radioactive 'tracers' deliberately introduced into ground water and to estimate health risks from natural and artificial radiation sources.

## 4.1 Natural radiation

Spontaneous radioactive decay produces alpha, beta and gamma rays. Alpha and beta 'rays' are actually particles; gamma-rays are a high-energy form of electromagnetic radiation which, quantum mechanics tells us, can also be treated as if composed of particles.

### 4.1.1 Alpha particles

An alpha particle consists of two protons held together by two neutrons to form the stable helium nucleus. Emission of alpha particles is the main process in radioactive decay, resulting in a decrease of four in atomic mass and two in atomic number. The particles have quite large kinetic energies but are rapidly slowed down by collisions with other atomic nuclei; at 'thermal' energies they soon gain two orbital electrons and become indistinguishable from other helium atoms. The average distance travelled in solid rock before this occurs is measured in fractions of a millimetre.

### 4.1.2 Beta particles

Beta particles are electrons ejected from atomic nuclei. They differ from other electrons only in having higher kinetic energies and so cease to be identifiable after being slowed down by multiple collisions. Energy is lost most rapidly in collisions with other electrons; in solids or liquids the average range of a beta particle is measured in centimetres.

## 4.1.3 Gamma radiation

Gamma 'rays' are electromagnetic waves with frequencies in excess of about $10^{16}$ Hz. Such radiation is more easily treated as consisting of photons characterized by energies proportional to the frequencies. The energy range for gamma rays is generally considered to start at about 0.1 MeV.

Because they are electrically neutral, photons interact with matter much less than do either alpha or beta particles and are consequently the principal form of radiation detected in geophysical surveys. Even so, approximately 90% of the radiation from a rock mass of density 2700 kg·m$^{-3}$ will come from within

20–30 cm of the surface and even for soil of density 1500 kg·m$^{-3}$ only 10% of the radiation will come from below about 50 cm. On the other hand, 100 m of air will absorb only about half of a gamma-ray flux, so that atmospheric absorption can generally be ignored in ground surveys.

## 4.1.4 Radioactivity of rocks

The geological usefulness of any form of radiation lies in its relationship to the decay of a specific unstable atomic nucleus. Decay is statistical; the average number of decays in a given time will be directly proportional to the number of atoms of the unstable element present, and this number will therefore decrease exponentially (Fig. 4.1). Decays are characterized by the half-lives, the time taken for half the original atoms to break down, or by decay constants. The two multiplied

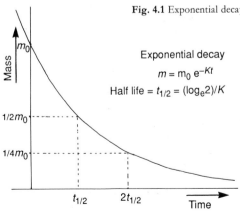

**Fig. 4.1** Exponential decay curve.

Exponential decay

$$m = m_0 \, e^{-Kt}$$

Half life $= t_{1/2} = (\log_e 2)/K$

together equal $\log_e 2$ ($=0.69315$), so that either quantity can be obtained by dividing this by the other.

Some elements found in nature have very short half-lives but exist because they are formed in decay series which originate with very long-lived isotopes, sometimes termed *primeval*. The principal primeval isotopes are $^{40}K$, $^{232}Th$, $^{235}U$ and $^{238}U$. Others, such as $^{48}Ca$, $^{50}V$ and $^{58}Ni$, exist but are either too rare or too weakly radioactive to be significant. Radioelements are mainly concentrated in acid igneous rocks and in sediments derived from them.

### 4.1.5 Radioactive decay series

The main radioactive decay schemes are shown in Table 4.1. $^{40}K$, which forms about 0.0118% of naturally-occurring potassium, decays in a single stage, either by beta emission to form $^{40}Ca$, or by electron capture (K-capture) to form $^{40}Ar$. The argon nucleus is left in an excited state from which it decays with the emission of a 1.46-MeV photon. The half-life of $^{40}K$ is 1470 Ma for beta decay and 11 800 Ma for K-capture.

The other important primeval radioisotopes decay into nuclei which are themselves unstable. As with $^{40}K$, there may be more than one possible decay mode, and the decay chains are quite complex. All, however, end in stable isotopes of lead. Of the two uranium primevals, only the decay series for $^{238}U$ is shown in Table 4.1. $^{235}U$ makes up only 0.7114% of the naturally-occur-ring element and, although its rather greater activity allows it to contribute nearly 5% of the overall uranium activity, it may for most practical purposes be ignored.

By no means all decay events are accompanied by significant gamma emission. The first stage in the decay of $^{232}Th$ involves only weak gamma activity and the strongest radiation in the chain comes from the decay of $^{208}Tl$, near the end. $^{208}Tl$ decay is accompanied by emission of a 2.615-MeV photon, the highest energy radiation to come from a terrestrial source.

In the $^{238}U$ chain, $^{214}Bi$ is notable for the numbers and energies of the gamma photons produced. Those at 1.76 MeV are taken as diagnostic of the presence of uranium.

The radon isotope, $^{222}Rn$, which precedes $^{214}Bi$ in the chain, has a half-life of nearly four days and so can disperse quite widely away from a primary uranium source. Gaseous dispersion has much less effect in thorium decay since the radon isotope formed, $^{220}Rn$, has a half-life of less than a minute.

### 4.1.6 Radioactive equilibria

If a large amount of a primeval isotope is present and if all the daughter products remain where they are formed, an equilibrium will eventually be established in which the number of atoms of each member created in a given time equals the number decaying. Only the concentrations of the two end members

**Table 4.1** Natural radioactive decay

| Parent | Mode | Daughter | Half-life | $\gamma$ energy (MeV) and % yield[1] |
|---|---|---|---|---|
| $^{238}$U | $\alpha$ | $^{234}$Th | $4.5 \times 10^9$yr | 0.09(15) 0.6(7) 0.3(7) |
| $^{234}$Th | $\beta$ | $^{234}$Pa | 24.1 day | 1.01(2) 0.77(1) 0.04(3) |
| $^{234}$Pa | $\beta$ | $^{234}$U | 1.18 min | 0.05(28) |
| $^{234}$U | $\alpha$ | $^{230}$Th | $2.6 \times 10^5$yr | |
| $^{230}$Th | $\alpha$ | $^{226}$Ra | $8 \times 10^4$yr | |
| $^{226}$Ra | $\alpha$ | $^{222}$Rn | 1600 yr | 0.19(4) |
| $^{222}$Rn | $\alpha$ | $^{218}$Po | 3.82 day | |
| $^{218}$Po | $\alpha$ | $^{214}$Pb | 3.05 min | |
| $^{214}$Pb | $\beta$ | $^{214}$Bi | 26.8 min | 0.35(44) 0.24(11) 0.29(24) 0.05(2) |
| $^{214}$Bi | $\beta$ | $^{214}$Po | 17.9 min | 2.43(2) 2.20(6) 1.76(19) 1.38(7) 1.24(7)$*$ |
| $^{214}$Po | $\alpha$ | $^{210}$Pb | $1.6 \times 10^{-4}$ sec | |
| $^{210}$Pb | $\beta$ | $^{210}$Bi | 19.4 yr | |
| $^{210}$Bi | $\beta$ | $^{210}$Po | 5.0 day | 0.04(4) |
| $^{210}$Po | $\alpha$ | $^{206}$Pb | 138.4 day | |
| $^{232}$Th | $\alpha$ | $^{228}$Ra | $1.4 \times 10^{10}$ yr | 0.06(24) |
| $^{228}$Ra | $\beta$ | $^{228}$Ac | 6.7 yr | |
| $^{228}$Ac | $\beta$ | $^{228}$Th | 6.1 hr | 1.64(13) 1.59(12) 0.99(25) 0.97(18) 0.34(11)$*$ |
| $^{228}$Th | $\alpha$ | $^{224}$Ra | 1.9 yr | |
| $^{224}$Ra | $\alpha$ | $^{220}$Rn | 3.64 day | |
| $^{220}$Rn | $\alpha$ | $^{216}$Po | 54.5 sec | |
| $^{216}$Po | $\alpha$ | $^{212}$Pb | 0.16 sec | |
| $^{212}$Pb | $\beta$ | $^{212}$Bi | 10.6 hr | 0.30(5) 0.24(82) 0.18(1) 0.12(2)$*$ |
| $^{212}$Bi | $\beta$(66%) | $^{212}$Po | 40 min | 1.18(1) 0.83(8) 0.73(10) |
| | $\alpha$(34%) | $^{208}$Tl | 97.3 min | |
| $^{212}$Po | $\alpha$ | $^{208}$Pb | $0.3 \times 10^{-6}$ sec | 2.62(100)  0.86(14)  0.58(83) |
| $^{208}$Tl | $\beta$ | | 3.1 min | 0.51(25)$*$ |
| $^{40}$K | $\beta$(89%) | $^{40}$Ca | $1.45 \times 10^9$ yr | |
| | K(11%) | $^{40}$Ar | $1.17 \times 10^{10}$ yr | 1.46(11) |

*Notes*: 1. Percent yields (shown in parentheses) indicate number, out of each 100 decays, producing photons of energy specified. Some single decay events produce more than one photon.
2. Decay branches which involve less than 10% of a parent element are not shown.
3. Photons of numerous other energies emitted in events marked $*$.

of the series change. An (imaginary) three-member decay chain beginning with a primeval radioelement A can be used as an example:

The rate of decay of A will be effectively constant over a period of tens or even thousands of years, and thus the same amount of the daughter element B will be produced every minute. The amount of B decaying per minute depends on the amount present; if there is initially very little, there will be few decays and the concentration of B will gradually increase.

If B is initially plentiful, more will decay than will be replaced by decay of A, and the concentration will decrease. Somewhere between the two extremes is a concentration of B which will give the same number of decays per minute for B as for A. The amount of A decreases and the amount of the end member C increases correspondingly, with no change in the amount of B. The same argument can be applied to decay chains of any length but equilibrium will not be established if gaseous or soluble intermediate products have half-lives long enough to allow them to be transported before they decay. The exhalation of radon by uranium ores notably disrupts equilibrium and the primary source of a 'uranium' anomaly may be hard to find.

Young deposits of radioactive minerals may not have reached equilibrium. Given the very long half-lives of the primeval radioelements, it is not surprising that equilibrium is generally not established in Tertiary or Quaternary deposits, and 'roll-front' uranium ores are notorious for the separation of the uranium concentrations from the zones of peak radioactivity.

### 4.1.7 *The natural gamma-ray spectrum*

Natural gamma rays range from cosmic radiation with energies above 3 MeV down to X-rays. A typical measured spectrum is shown in Fig. 4.2. The individual peaks correspond to specific decay events, the energy of each photon falling somewhere in a small range determined by the nuclear kinetic energies prior to decay and by errors in measurement.

The background curve upon which the peaks are superimposed is due to 'scattered' cosmic (mainly solar) and terrestrial radiation. Gamma rays can be scattered in three ways. Very energetic cosmic photons passing close to atomic nuclei may form electron–positron pairs. The positrons soon interact with other electrons, producing more gamma rays. At lower energies, a gamma ray may eject a bound electron from an atom, some of the energy being transferred to the electron and the remainder proceeding as a lower-energy photon (*Compton scattering*). Alternatively, a photon may be totally absorbed in ejecting a bound electron from an atom (*photoelectric* effect).

## 4.2 Radiation detectors

The earliest radiometric instruments

relied on the ability of radiation to ionise low-pressure gas and initiate electrical discharges between electrodes maintained at high potential differences. These Geiger–Müller counters are now considered obsolete. They respond mainly to alpha particles and suffer long 'dead' periods after each count, during which no new events can be detected.

Gamma rays produce flashes of light when they are photoelectrically absorbed in sodium iodide crystals. Small amounts of thallium are added to the crystals, which are said to be 'thallium-activated'. The light can be detected by photomultiplier tubes (PMTs) which convert the energy into electric current. The whole sequence occupies only a few microseconds and events overlap only at very high count rates.

A scintillometer consists of a crystal, one or more PMTs, a power

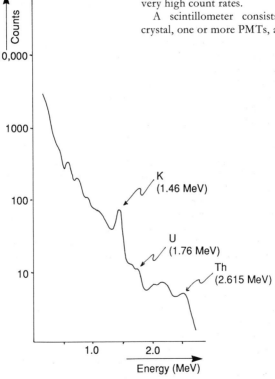

Fig. 4.2 A natural gamma-ray spectrum.

65

supply (which must provide several hundred volts for the PMTs), and some counting circuitry. The results may be displayed digitally but quite commonly are shown by a needle on an analogue 'rate-meter'. Some instruments produce an audible 'click' each time a gamma ray is detected, and have alarms which can be triggered when the count rate exceeds a predetermined threshold, so that the dial need not be continually observed.

Where a digital display is used, a count time must be selected by the operator, but rate-meter averaging is controlled by a time constant. If this is too short, the needle will be in continual motion and readings will be difficult to take. With too long a time constant, the response is slow and narrow anomalies may be overlooked.

The sensitivity of a scintillometer depends almost entirely on crystal size, a bigger crystal recording more events. Count rates are thus not absolute but depend on the instrument and crystal used. Instruments of the same type and model should read roughly the same in the same places, but even this needs to be checked carefully since radioactive contaminants near, and within, the crystals can cause differences in readings.

Different models of scintillometer are likely to record very different count rates, one reason being that crystals may be shielded so as to detect radiation from one direction only. If it is essential that comparable data are obtained, portable radio-active sources can be used for calibration and also to check the extent of shielding. Even instruments of the same type may have surprisingly different apertures.

Standard sources provide strictly valid comparisons for one particular gamma-ray energy only, and calibration sites have been established in several countries where instrument performance can be checked over concrete pads containing carefully controlled concentrations of various radioelements. However, it is best, if several instruments are to be used, to make comparative checks in the actual survey area before attempting to reduce all results to common equivalent readings. Bases at which this has been done should be described for the benefit of later workers.

The lack of absolute standards is less significant in radiometric work than it would be in other branches of applied geophysics, since the observations are of a random process and even successive readings with the same instrument are unlikely to agree unless counts are averaged over very long periods of time.

*4.2.2 Gamma-ray spectrometers*

If a pulse-height analyzer is incorporated into the PMT circuitry, the energy of each gamma photon which produces a scintillation event can be estimated. Events with energies within certain pre-determined energy ranges ('windows') or above pre-selected energy 'thresholds' can be counted separately. The entire gamma-ray flux can be observed at a series of narrow adjoining windows to obtain a curve similar to that shown in Fig. 4.2.

Strictly, the term 'spectrometer' should be reserved for those instruments, with perhaps 256 or more channels, which can record a complete spectrum, but in practice it is applied to any multi-channel instrument with some degree of energy discrimination. Usually there are only four channels, one for total count and one each for the $^{208}$Tl peak at 2.62 MeV, $^{214}$Bi at 1.76 MeV and $^{40}$K at 1.46 MeV. Typical 'windows' through which these peaks could be observed might extend from 2.42 to 2.82 MeV, from 1.66 to 1.86 MeV and from 1.36 to 1.56 MeV respectively. Concentrations of all three parent elements can thus be estimated, although care is needed in recognizing disequilibrium.

Many spectrometers are designed to be compatible with a number of different crystals; the appropriate crystal is selected on the basis of cost, of time available for survey work and of the accuracy required.

### 4.2.3 Alpha-particle monitors

Useful radiometric data may be obtained by monitoring alpha activity, and hence the radon content of the soil gas, even in areas where no obvious gamma-ray anomalies can be found. Nearby uranium mineralization can be indicated, although radon diffuses so readily through rocks and soil that locating the primary source may be difficult.

One of the commonest forms of monitor is a thin metallic membrane mounted in a card frame about the size of a 35-mm slide, on which radon daughter products can collect. The card is suspended in an inverted flower pot or similar container in a hole about 0.5 m deep.

The hole is covered with a plastic sheet and earth is then poured on top until the ground surface is again level (Fig. 4.3). The sheet should be sufficiently large for its edges to project, so that it can be lifted out together with the overlying soil when the card is to be removed. The hole is left covered for at least 12 hours, which is long enough for equilibrium to be established. Longer periods of burial will not alter the reading.

After the card has been removed it is placed in a special 'reader' which is sensitive to alpha radiation. If an area is systematically investigated, anomalies too subtle to appear on a conventional radiometric map may be located. Other types of alpha detector are available which have the reading and data storage elements included in the field apparatus.

## 4.3 Radiometric surveys

Ground radiometric surveys tend to be rather frustrating operations. Because of the shielding effect of even thin layers of rock or soil, it is very easy to overlook concentrations of radioactive minerals in rocks which are only patchily exposed at

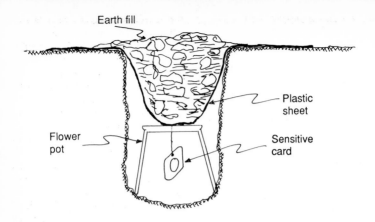

**Fig. 4.3** Alpha-radiation card in its hole.

the surface. Reliance on stations placed at uniform distances along a traverse may be unwise, and the field observer needs to be more than usually aware of his environment.

### 4.3.1 Reading times

Accurate radiometric data can be obtained only by occupying each station long enough for the statistical variations in the count to average out. What this implies will depend on the count levels themselves and must be determined by actual experiment. The percent statistical error is sometimes considered to equal one hundred divided by the square root of the number of counts, and so is about 30% for 10 counts and only 1% for 10 000. A period of time which is fully adequate for total count readings may be insufficient

for the readings on the K, U and Th channels.

It may be that areas where count rates are low are in any case of little interest, and there is then little point in wasting time obtaining accurate data. In these circumstances it may be sufficient to cover ground at a slow walk, using a rate-meter. The rate of progress should be such that the narrowest source of interest would not be crossed completely in a time equal to the time constant of the instrument.

Even when a spectrometer is used, it is usual to record only total count in the first instance, reserving the more time-consuming 'spectral' readings for areas of total-count anomaly. There are, of course, dangers in this approach, as the concentration of one radioelement may decrease in the same region as another increases.

### 4.3.2 Stripping ratios

To estimate thorium, uranium and potassium abundances from spectrometer readings, corrections must be made for gamma rays scattered from other parts of the spectrum. The thorium peak must be corrected for cosmic radiation and for the 2.43 MeV radiation from $^{214}$Bi in the uranium decay chain, which overlaps into the commonly used 'thorium' window. The uranium count must be corrected for thorium and the potassium count for both elements. The correction process is known as *stripping*.

Stripping factors are generally listed in equipment manuals and in some cases can be applied by built-in circuitry so that the corrected results are displayed directly. The factors vary from detector to detector, primarily with variations in crystal size. It is always assumed that the decay series observed are in equilibrium and interpretational errors will be made if this is not so. It is generally preferable to record actual count rates and 'strip' at the processing stage.

### 4.3.3 Radiometric assays

If a bare rock surface is available, a gamma-ray spectrometer can be used to make quantitative assays of thorium, uranium and potassium. The rock should be dry, so that absorption by moisture, either on or below the surface, is not a factor. Observations must be taken over a sufficiently long period for statistical fluctuations to be smoothed out. In practice this means accumulating at least 1000 counts; 10 000 would be needed for 1% accuracy.

Radioelement concentrations are determined by inserting the observed count rates into equations, provided by the manufacturers, which are specific to the instrument and crystal being used. If the material is so radioactive that rates exceed 1000 cps, corrections will be needed for 'dead' time. As a single count takes a few microseconds, at 10 000 cps the instrument will be dead for several tens of milliseconds in each second.

### 4.3.4 Geometrical considerations

Source 'geometry' is important in all radiometric surveys and especially in assay work. Radiation comes from a very thin layer at the ground surface and only weak anomalies can be expected if the lateral extent of the source is small compared with the distance to the detector. On the other hand, the height of the detector above an extended source should not greatly affect the count rate, merely altering the size of the area to which the measurement refers.

The optimum '$2\pi$' geometry is only rarely obtained in practice. Some other possible source geometries and factors for correction to standard $2\pi$ values are shown in Fig. 4.4.

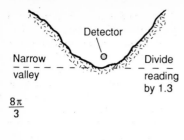

Narrow valley

Detector

Divide reading by 1.3

$\frac{8\pi}{3}$

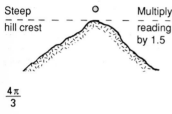

Steep hill crest

Multiply reading by 1.5

$\frac{4\pi}{3}$

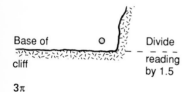

Base of cliff

Divide reading by 1.5

$3\pi$

**Fig. 4.4** Geometries and correction factors in radiometric surveys.

*4.3.5 Corrections for background variations*

Although often vital in airborne work, corrections for atmospheric radon and for cosmic background are usually less than 10% and unimportant in ground surveys. If thought necessary because very subtle variations are being observed or precise assay work is being done, they can be estimated by taking a reading either over the middle of a body of water 1-m deep and at least 10-m across or with the bottom of the detector shielded with lead sheet. Neither of these methods is likely to be very convenient. Background variations, mainly due to atmospheric humidity changes (wet air absorbs radiation far more efficiently than dry) can be monitored using an additional, fixed detector.

Some background radiation originates in radioactive material present in the detector crystal and housing. The level should be constant over long periods and might be measured by placing the detector in a totally shielded environment. In practice this is likely to be difficult to arrange. The correction is usually trivial, it being far more important to ensure that dirt, which might be contaminated, is not smeared on the detector housing.

A still more important possible source of spurious radiation is the observer himself. Watches with radioactive luminous dials are now rare, but compasses need to be carefully checked. Obviously, a calibration source should not be carried in the field. Radioactive contamination of the observer is especially significant if the sensor is carried in a back pack; in these circumstances the absorption of radiation by his body also has to be taken into account, usually by direct experiment.

A small amount of radioactive material is included in the vacuum chambers of quartz gravity meters to prevent build-ups of static electricity on the spring system. Radiometric and gravity surveys are occasionally done together, and absurd conclusions can be reached.

### 4.3.6 Recording radiometric data

Because gamma radiation is heavily absorbed by both rock and soil, comprehensive notes should be taken during radiometric surveys. Departures from '$2\pi$' geometry must be noted, together with details of the soil cover. If bare rock cannot be seen, some attempt should be made to decide whether the overburden developed *in situ* or was transported into place, and to estimate its thickness. Weather conditions can also be important; in particular, since absorption is much greater in wet than in dry soil, recent rain and the presence of puddles of standing water should always be noted.

The way in which readings are taken, including time constants or count periods, must be recorded. Field techniques should not be varied in the course of a survey and the location of the sensor (e.g. whether hand-held or in a back-pack) should be specified.

# 5

## *Electric current methods— general considerations*

A number of geophysical methods use measurements of voltages or magnetic fields associated with electric currents flowing in the ground. The currents may be natural but are more often artificially produced by direct contact or electromagnetic induction.

## 5.1 Resistivity

Graphite and most metallic sulphides conduct electricity quite efficiently by flow of electrons, but most rock-forming minerals are very poor conductors. Ground currents are therefore mainly carried by ions moving in pore waters.

### 5.1.1 Ohm's Law and resistivity

The current in a conductor is generally equal to the voltage across it divided by a constant, the resistance. This is *Ohm's Law*. Resistance ($R$) is measured in ohms when current ($I$) is in amps and voltage ($V$) is in volts. The resistance of a unit cube to current flowing between opposite faces is termed the resistivity ($\rho$) and the resistance of a rectangular block of material is proportional to the distance the current must flow ($x$) and inversely proportional to the cross-sectional area ($A$), i.e.

$$V = IR \tag{5.1}$$

and

$$R = \rho(x/A) \tag{5.2}$$

Resistivity is measured in ohm-metres. The reciprocal, conductivity, measured in mhos per metre, is also sometimes used.

Isotropic materials have the same resistivity in all directions. Most rocks are reasonably isotropic but strongly laminated slates and shales are more resistive across the laminations than parallel to them.

### 5.1.2 Electrical resistivities of rocks and minerals

The resistivity of a rock is roughly equal to the pore fluid resistivity divided by the fractional porosity. *Archie's Law* states that resistivity is

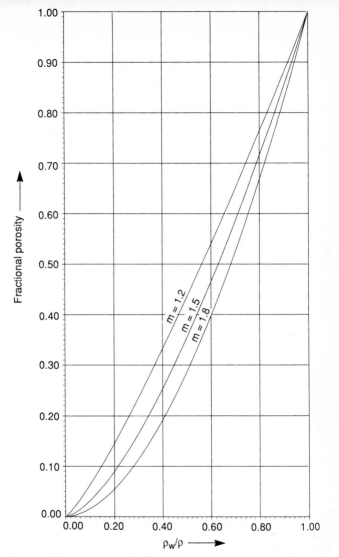

**Fig. 5.1** Archie's Law variation of bulk rock resistivity $\rho$, for rock with insulating matrix and pore-water resistivity $\rho_w$. $m$ is 1.2 for spherical grains and about 1.8 for platey or tabular ones.

proportional to fractional porosity raised to a power of between about 1.2 and 1.8, according to the shape of the matrix grains. The departures from linearity are not large for most common porosities (Fig. 5.1). Pure water is ionized to only a very small extent, so that conduction in pore waters depends on the presence of dissolved salts, mainly sodium chloride (Fig. 5.2). Clay minerals are ionically active and clays conduct well if even slightly moist.

Resistivities of common rocks and minerals are shown in Table 5.1. Resistivities of more than 10 000 ohm-metre or less than 10 ohm-metre are rarely encountered in field surveys.

ground resistivity by passing current between a pair of grounded electrodes and measuring the voltage between them does not work, because of contact resistances which may amount to thousands of ohms. The problem can be solved if voltage is measured across a second pair of electrodes, as long as virtually no current flows through them. A high-impedance voltmeter is thus a necessity and a geometric factor is needed to convert the readings obtained with the four-electrode 'array' to a resistivity.

Contact resistances at current electrodes limit current flow but do not affect resistivity calculations.

### 5.1.3 Contact resistances

The 'obvious' method of measuring

### 5.1.4 Apparent resistivity

Any single measurement with any

**Table 5.1** Resistivities of common rocks and ore minerals (ohm-metres)

| Common rocks | | | |
|---|---|---|---|
| Topsoil | 50–300 | Greenstone | 500–200 000 |
| Loose sand | 500–5000 | Gabbro | 1000–500 000 |
| Gravel | 100–6000 | Granite | 200–100 000 |
| Clay | 1–100 | Basalt | 50–200 000 |
| Weathered bedrock | 100–1000 | Graphitic schist | 10–500 |
| Sandstone | 200–8000 | Slates | 500–500 000 |
| Limestone | 500–10 000 | Quartzite | 500–800 000 |
| | | | |
| Ore minerals | | | |
| Pyrite (ores) | 100–0.01 | Pyrrhotite | 0.01–0.001 |
| Chalcopyrite | 0.1–0.005 | Galena | 100–0.001 |
| Sphalerite | 1 000 000–1000 | Cassiterite | 10 000–0.001 |
| Magnetite | 1000–0.01 | Hematite | 1 000 000–0.01 |

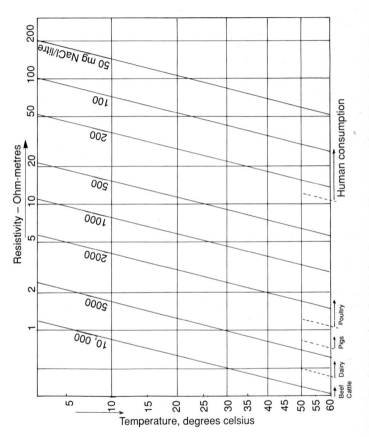

**Fig. 5.2** Variation of resistivity of water with concentrations of dissolved salt. The uses to which water of various qualities can be put are also shown.

array can be interpreted as due to homogeneous ground of some constant resistivity. This *apparent resistivity* is denoted by the symbol $\rho_a$. The geometric factors used to calculate apparent resistivity can be derived from first principles. The electrical potential, a distance $a$ from a point electrode at the surface of homogeneous ground (*jargon*: a uniform half-space) of resistivity $\rho$ is:

$$\rho I / 2\pi a \qquad (5.3)$$

where $I$ is the current, which may be positive (into the ground) or negative. The potential at any point due to the current flow is equal to the sum of the contributions from the individual current electrodes. The voltage measured in a four-electrode survey over homogeneous ground is given by

$$V = \frac{I\rho}{2\pi}\left(\frac{1}{R_m} - \frac{1}{r_m} - \frac{1}{R_n} + \frac{1}{r_n}\right) \qquad (5.4)$$

where $V$ is the voltage difference between electrodes m and n due to a current $I$ flowing in the ground. Distances from these electrodes to

the positive and negative current electrodes are indicated by $R$ and $r$ respectively, i.e. $R_m$ is the distance from voltage electrode m to the positive current electrode.

The factors are unaffected by interchanging current and voltage electrodes but voltage electrode spacings are normally kept small to minimize the effects of natural potentials.

## 5.2 Electrode arrays

Figure 5.3 shows the common electrode arrays and their geometric factors. The names are those of present field usage and may upset pedants. A dipole, for example, should consist of two voltage or two current electrodes separated by a distance which is negligible compared to the overall size of the array, and use of the term in the dipole–dipole and pole–dipole arrays, where $n$ is an integer between 1 and 6, is thus formally incorrect. Not many people worry about this.

A fixed electrode 'at infinity' should be at least ten times as far from any other electrode as the separation between any moving elec-

**Example 5.1**

---

Geometrical factor for the Wenner array (Fig. 5.3a):

$R_m = a \quad r_m = 2a \quad R_n = 2a \quad r_n = a$

$$V = \frac{I\rho}{2\pi a}\left(1 - \tfrac{1}{2} - \tfrac{1}{2} + 1\right) = \frac{I\rho}{2\pi a}$$

$$\rho = 2\pi a \, \frac{V}{I}$$

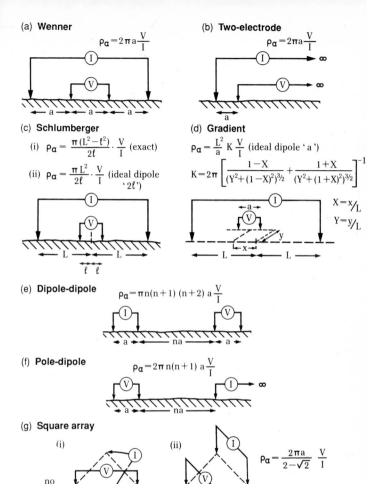

**Fig. 5.3** Some common electrode arrays and their geometric factors. (*a*) Wenner; (*b*) Two-electrode; (*c*) Schlumberger; (*d*) Gradient; (*e*) Dipole–dipole (Eltran); (*f*) Pole–dipole; (*g*) Square array [(*i*) diagonal (*ii*) broadside]. There is no geometric factor for the diagonal square array, as no voltage difference will be observed over homogeneous ground.

trodes. Field work with such systems can be hampered by the long cables, which are also vulnerable to inductive noise.

## 5.2.1 Array descriptions (Fig. 5.3)

*Wenner Array*: Very widely used, with a vast amount of interpretational literature. The common 'compromise' array against which others are assessed.

*Two-Electrode Array*: Of some theoretical interest since it is possible to calculate from the readings the results which would be obtained from any other type of array, providing coverage is adequate. 'Noise' and lack of precision prevent much use being made of this fact and the array is little used in surface work because of practical difficulties. As the *normal* array it is one of the most popular systems in electrical well-logging.

*Schlumberger Array*: The only array to rival the Wenner in availability of interpretational material, all of which relates to the 'ideal' array with negligible distance between the inner electrodes. Favoured, along with the Wenner, for electrical depth-sounding work.

*Gradient Array*: Used principally in reconnaissance work. Large numbers of readings can be taken on parallel traverses without moving the current electrodes. Powerful generators are usually needed. Figure 5.4 shows how the geometrical factor given in Fig. 5.3d varies with the position of the voltage dipole.

*Dipole–Dipole (Eltran) Array*: Popular in induced polarization (IP) work because of the complete separation of current and voltage circuits. A considerable body of interpretational literature available. Information from different depths obtained by changing $n$, i.e. by looking at parts of the current flow which have penetrated more or less deeply. Results usually plotted as pseudo-sections (Section 7.4.2).

*Pole–Dipole Array*: Depth penetration is varied by varying $n$; results can be plotted as pseudo-sections. The asymmetric anomalies are more difficult to interpret than those produced by symmetric arrays. Peaks are displaced from the centres of conductive or chargeable bodies, and there is no real agreement as to where the results should be plotted. A point mid-way between the moving voltage electrodes is most commonly chosen. Electrode positions have to be recorded with especial care.

*Square Array*: Four electrodes are positioned at the corners of a square and are variously combined into voltage and current pairs (Fig. 5.3g). Depth soundings are made by expanding the square; in traversing the entire array is moved laterally. Inconvenient, but can provide an experienced interpreter with vital information about ground anisotropy and inhomogeneity. Few published case histories or type curves.

*Multi-Electrode Arrays (not shown)*: The *Lee array* resembles the Wenner array but has an additional central electrode. The voltage differences from the centre to the two 'normal'

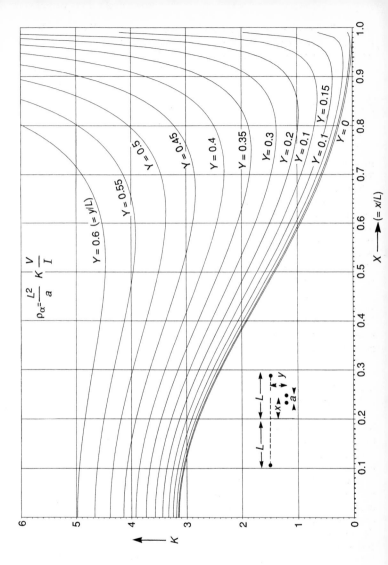

**Fig. 5.4** Variation in gradient array geometric factor with distance along and across line, for array parameters as shown. Array total length 2L, voltage dipole spacing a.

voltage electrodes give a measure of ground inhomogeneity. The two values can be summed for application of the Wenner formula. Other multi-electrode arrays are designed to focus current into the ground to give deep penetration without large expansion. These are controversial and should be used only under the guidance of an experienced interpreter.

### 5.2.2 Signal-contribution sections

Current-flow patterns in a one- and two-layered Earth are shown in Fig. 5.5. The effects of near-surface inhomogeneities, which strongly influence the choice of array, are graphically illustrated by signal-contribution sections (Fig. 5.6). These are contour plots of the contribution made by each unit volume of ground to the measured voltage, and hence to the apparent resistivity. For linear arrays the contours have the same appearance in any plane, whether vertical, horizontal or dipping, through the line of electrodes.

A reasonable first reaction to Fig. 5.6 is that useful resistivity surveys are impossible, as the contributions from ground close to the electrodes are very large. Some disillusioned clients might endorse this view. However, the variations in sign imply that a conductive body in some places reduces and in other places increases the apparent resistivity, and in homogeneous ground the near-electrode effects cancel quite precisely.

When a Wenner or dipole–dipole array is expanded, all the electrodes are moved and the contributions from near-surface bodies vary from reading to reading. With a Schlumberger array, near-surface effects vary much less, provided that only the outer electrodes are moved, and for this reason the array is often preferred for depth sounding. However, offset techniques (Section 6.3.3) allow excellent results to be obtained with the Wenner.

Near-surface effects may be large when a gradient array is used for profiling but are also very local; a smoothing filter can be used.

### 5.2.3 Depth penetration

The signal-contribution contours for the Wenner array at depth are slightly flatter than those for the Schlumberger array, suggesting that the Wenner will locate flat-lying interfaces more accurately. The contours for the dipole–dipole array are near vertical in some places at considerable depths, showing that the array is best suited to mapping lateral changes. Arrays are usually chosen at least partly for their depth penetration, which is very hard to define.

The depth to which a given fraction of current penetrates in a layered Earth depends on the layering as well as on the separation between the current electrodes. Voltage electrode positions determine only which part of the current field is sampled, and the depth penetration of Wenner and Schlumberger arrays is thus likely to

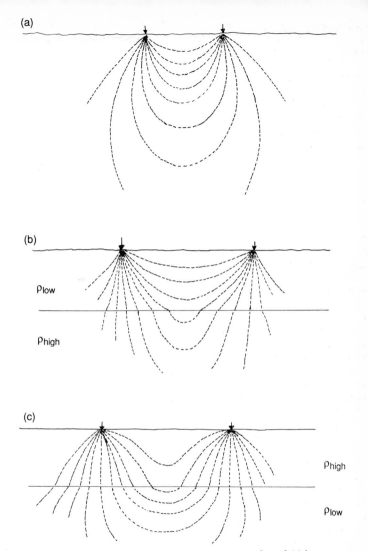

**Fig. 5.5** Current flow patterns between electrodes at the surface of: (*a*) homogeneous ground; (*b*) two-layer ground, lower resistivity near surface; (*c*) two-layer ground, higher resistivity near surface.

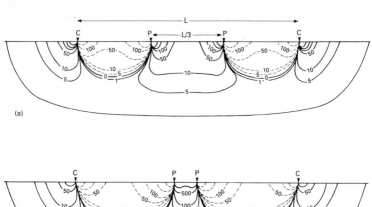

(a)

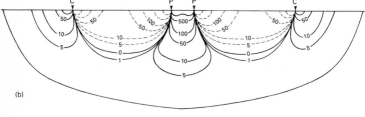

(b)

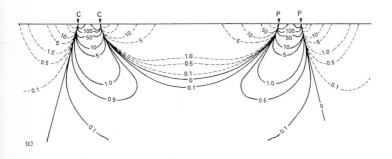

(c)

**Fig. 5.6** Signal contribution sections for (*a*) Wenner (*b*) Schlumberger and (*c*) dipole–dipole arrays. Contours show the relative contributions to the signal from unit volumes of homogeneous ground. Dashed lines indicate negative values. (Reproduced courtesy Dr. R. Barker.)

be very similar for similar total array lengths. This is confirmed by the similarity in the two-layer type curves (see Fig. 6.3). The point at which the effect of a deep interface becomes evident depends on the resistivity contrast but is of the order of a quarter of the current-electrode spacing. This is true also of the gradient array, which is a generalized version of the Schlumberger array but which is not used in the same way to define horizontal layers.

For any array there is a depth at which the effect of a thin layer of different resistivity in otherwise homogeneous ground will be a maximum. This would be expected to be less than the depth at which the interface in a two-layer Earth first becomes apparent, and plots of the effects of such a layer, shown in Fig. 5.7 for the Wenner, Schlumberger and dipole–dipole arrays, confirm this. By this criterion the Wenner is the least, and the dipole–dipole the most, penetrative. However, the Wenner curve is the most sharply peaked, confirming its excellent resolving power.

### 5.2.4 Conductive overburden

In many arid tropical areas, the build-up of salts in the soil leads to the development of near-surface layers of very high conductivity, which effectively short-circuit current flow and allow very little to penetrate to deeper levels (see Fig. 5.5b).

Some arrays are more susceptible to conductive overburden than others. The dipole–dipole array, which in any case measures the lowest voltage of any common array for a given depth of investigation, is

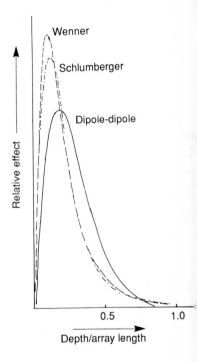

**Fig. 5.7** Relative effects of a thin, horizontal high-resistance bed in otherwise homogeneous ground. The depth penetration of the Wenner array is very slightly less than that of the Schlumberger array. The areas under all three curves have been made equal, concealing the fact that, for a given input current, a smaller voltage will be observed using the Schlumberger array and a very much smaller voltage will be observed using the dipole–dipole array.

particularly vulnerable. When using gradient, two-electrode or pole–dipole arrays, it is worth spending considerable time and effort to locate sites for the remote fixed electrodes where the conductive layer is thin or absent. Man-made conductors such as steel piping, wire fences and metallic scrap are usually unimportant if current is applied directly to the ground and only resistivity is being measured. They have to be very large in relation to array size, and very efficiently coupled to the electrodes, to compete with the low resistance presented by the ground.

## 5.3 Equipment for resistivity surveys

Resistivity and IP surveys require instruments and some means of making electrical contact with the ground. Cables and electrodes are cheap but vital parts of any system, and it is with these that much of the background 'noise' is associated.

### 5.3.1 Metal electrodes

Current electrodes are nearly always metal stakes. The contact resistance, which is the major limitation on current flow, depends on ground moisture and contact area. In dry ground the stakes may need to be hammered in to depths of more than half a metre and be watered to improve contact. Where contact is very poor, salt water and multiple stakes may be used.

Metal electrodes come in many different forms. Lengths of drill steel are excellent if the ground is stony and the stakes will be heavily hammered. Pointed lengths of angle-iron are only slightly less robust and have larger contact areas. If the main consideration is speed, large numbers of metal tent pegs can be pushed in along the line by an advance party. Stainless steel is best for potential electrodes, since it polarizes less than most other metals.

### 5.3.2 Non-polarizing electrodes

A metallic conductor in contact with the ground water, or with any electrolyte other than a saturated solution of one of its own salts, generates a contact potential. Non-polarizing electrodes can be used if such potentials are a serious problem. Most of these consist of copper rods in contact with saturated solutions of copper sulphate. The rod is attached to the lid of a container or 'pot' with a porous base of wood or, more commonly, unglazed earthenware (Fig. 5.8). Contact with the ground is made via the solution which leaks through the base. Some solid copper sulphate should be kept in the pot to ensure saturation. The temptation to 'top up' with fresh water must be resisted, as potentials will be generated if any part of the solution is less than saturated.

The high resistance of copper–copper sulphate electrodes is unimportant in voltage-measuring circuits in which currents do not flow. It may sometimes be desirable to use non-polarizing current electrodes in IP work but not only does resistance then become a problem, but the electrodes deteriorate rapidly due to electrolytic plating-out or dissolution of copper.

Copper sulphate solution gets everywhere and rots everything, and non-polarizing electrodes are not popular. They are essential in IP and SP work, but in ordinary resistivity surveys the simpler method of frequent current reversal is adopted. This also reduces the influence of natural fields, which then only cause serious problems if, as is only likely if the voltage electrodes are far apart, they are large compared to the potentials being measured.

### 5.3.3 Cables

Cables used in resistivity work are normally single core, multi-strand copper wires insulated with PVC. Thickness is usually dictated by the need for mechanical strength rather than low resistance, since the resistance at the ground contacts will always be very much higher than in the cable. Steel reinforcement may be needed for long lines.

Virtually all electrical work involves four cables, at least two of them long. Good practice in cable handling (Section 1.2.2) is essential if delays are to be avoided. Cables are usually connected to electrodes by crocodile clips; screw connections are slow and impractical, and are

**Fig. 5.8** Porcelain non-polarizing electrodes designed to be pushed into a shallow scraping made by a boot-heel on the ground surface. Other types are made to be pushed into a small hole made in the ground with a geological pick.

easily damaged by careless hammer blows. Clips are easily lost and everyone on the crew should carry at least one spare, a screwdriver and a small pair of pliers.

In all electrical surveys and especially in IP work, electromagnetic induction between current and voltage cables must be avoided. This is most easily arranged using dipole–dipole arrays and is the main reason for their popularity in IP. Precautions are also needed against induction from sources unconnected with the survey. Power lines often cause noise in IP surveys and can affect results even in 'd.c.' work. Power frequencies (50 or 60 Hz) are outside the 'd.c.' range but can give rise to interference, particularly with the more sophisticated modern instruments. In resistivity work, the results are likely to be absurd or non-existent, rather than misleading, but in IP apparently valid results can be produced.

### 5.3.4 Generators and transmitters

The apparatus which controls and measures current in a resistivity or IP survey is usually known as the *transmitter*. To minimize natural current and electrode-polarization effects, the direction of current flow is generally reversed, with a cycle time of between 0.5 and 2 seconds. IP waveforms are more complicated and must be very precisely controlled; primary current must be reduced to zero from its steady-state value within a few milliseconds.

The power source connected to a transmitter may be a dry or rechargeable battery or a motor generator. Hand-cranked generators have been used, notably in 'Megger' instruments, but are now rare. Generators with several kVA output are needed if current electrodes are more than one or two hundred metres apart. These are not very portable and supply power at levels which can be lethal. Stringent safety precautions must be observed, not only in handling the electrodes but also in monitoring the whole lengths of the current cables for passers-by and livestock.

### 5.3.5 Receivers and detectors

The voltage-measuring units in resistivity and IP surveys are sometimes known as *receivers*. In resistivity work their only function is to measure a voltage while drawing minimal current. High-sensitivity moving-coil instruments or potentiometric (voltage balancing) circuits were originally used, but have been almost entirely replaced by units based on field-effect transistors (FETs).

Transmitters and receivers in many modern low-power instruments are contained in single housings and readings are displayed, usually digitally, directly in ohmmetres. Current levels still have to be monitored, since low currents may affect the validity of the resistivity results. Often one of a number of pre-set levels is selected, and an error is indicated if this level cannot be maintained. It should also be possible to measure voltage. Most commercial instruments allow this, and can therefore be used for surveys of natural ground potentials (SP).

Digital readings are usually obtained over a fixed number of square-wave cycles. The number used represents a compromise between speed of coverage and good signal to noise ratio. Normally a reading is displayed as each cycle is completed, and the number of cycles used should be sufficient to allow this reading to stabilize.

Induced-polarization receivers must measure integrated or instantaneous voltages, or voltages at different frequencies (Chapter 7). They can also be used for resistivity and often for SP surveys, since steady-state voltages are measured

and SP voltages must be 'backed-off', but they may not be very convenient for these purposes.

### 5.3.6 Microprocessor control

Most modern combined transmitter–receivers are controlled by a microprocessor. Current levels, voltage settings, cycle periods, numbers of cycles and read-out formats can be programmed via front-panel keypads or switches. Low current, low voltage, incorrect or missing connections and other faults are all recognized automatically and are usually indicated by a numerical code which is meaningless without the handbook. If all else fails, read the instructions.

# Direct-current methods

Natural currents are always present in the Earth and the associated self- or spontaneous potentials (SPs) can be measured. Currents can also be artificially introduced; in 'd.c.' surveys the flow is either in a single direction or is reversed at intervals of one or two seconds.

## 6.1 SP surveys

SP surveys were at one time popular in mineral exploration because of their cheapness and simplicity. They are now relatively little used because some near-surface ore bodies which are readily detected by other electrical means give no SP anomaly.

### 6.1.1 Origins of natural potentials

Natural potentials of as much as 1.8 V have been observed where alunite is weathering to sulphuric acid, but the negative anomalies produced by sulphide ore bodies or graphite are generally less than 500 mV. The conductor should extend from the zone of oxidation near the surface to the reducing environment below the water table, thus providing a low-resistance path for oxidation–reduction currents (Fig. 6.1).

Small potentials, seldom exceeding 100 mV and usually very much less, may accompany groundwater flow. The sign of the effect depends on the rock mineralogy and on the mobilities and chemical properties of the ions in the pore waters but most commonly the region towards which groundwater is flowing becomes more electro-positive than the source area. These 'streaming potentials' are sometimes useful in hydrogeology but are noise in mineral exploration surveys, which may be inadvisable for up to a week after heavy rain.

Movements of steam or hot water can also explain most of the SPs associated with geothermal systems, but small ($< 10$ mV) voltages, which may be positive or negative, are produced directly by temperature differences. Geothermal SP anomalies tend to be broad (perhaps several kilometres across) and have

**Fig. 6.1** Sources of SP effects. The sulphide mass straddling the water-table concentrates the flow of oxidation–reduction return currents, producing a negative anomaly at the surface. The downslope flow of groundwater after rain produces a temporary SP, in this case inversely correlated with topography.

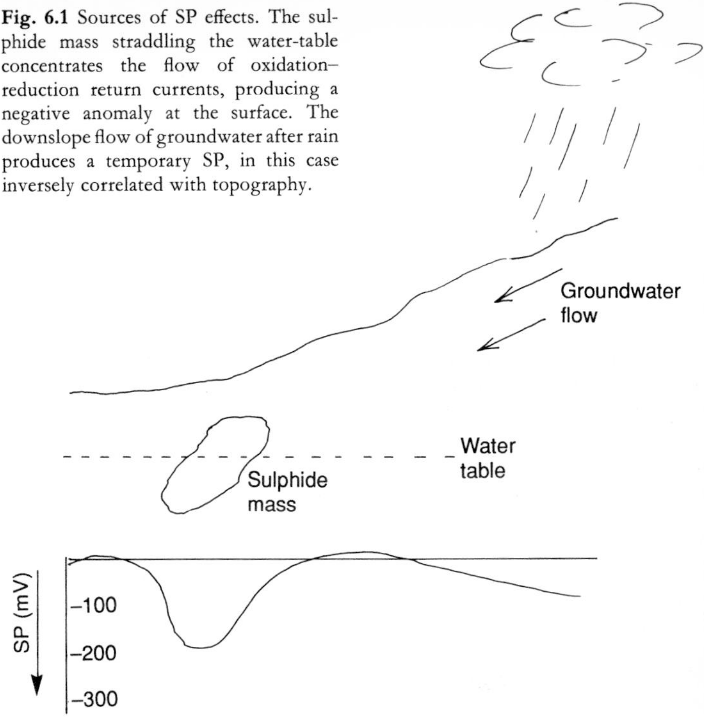

amplitudes of less than 100 mV, so very high accuracy is needed.

Small alternating currents are induced by variations in the Earth's magnetic field and by thunderstorms. Only the long-period components of the associated voltages, seldom amounting to more than 5 mV, will be detected by the d.c. voltmeters used in SP surveys. If, as is very occasionally the case, such voltages are significant, the survey should be repeated at different times of day so that the results can be averaged.

### 6.1.2 SP fieldwork

Voltmeters used for SP work must read to a millivolt and have very high impedances so that the currents drawn from the ground are negligible. Copper–copper sulphate 'pot' electrodes (Section 5.3.2) are almost universal, linked to the meters by lengths of insulated copper wire.

An SP survey can be carried out by traversing with two electrodes separated by a small constant distance, commonly 5 or 10 m, to measure average field gradient. The method is useful if cable is limited, but errors tend to accumulate and coverage is slow because the voltmeter and both electrodes must be moved for each reading.

More commonly, voltages are measured in relation to a single fixed base. One electrode and the voltmeter remain at this point and only the 'field' electrode is moved. A secondary base must be established before the cable runs out or the separation becomes too great for easy communication. Voltages measured from the new and old bases can be related to each other provided that the potential difference between the two is accurately known. This is usually measured both directly and indirectly.

Figure 6.2 shows how a secondary base can be established. At field point B the end of the cable has almost been reached but it is still possible to obtain a reading at the next point, C, using the original base at A. After differences have been measured between A and both B and C, the field electrode is left at C and the base electrode is moved to B.

The potential difference between A and B is thus determined both by direct measurement and by sub-tracting the B to C potential from the directly-measured A to C potential. The average can be added to values obtained with the base at B to obtain values relative to A.

### 6.1.3 Errors and precautions

If two estimates of a base/sub-base difference disagree by more than one or two millivolts, something is seriously wrong and work should be stopped until the reason is known. Usually it will be found that copper sulphate solution has either leaked away or become undersaturated. Pots should be checked every two to three hours by placing them on the ground a few inches apart; the voltage difference should not exceed 1 or 2 mV.

The accumulation of errors in large surveys can be minimized by working in a series of closed and inter-connecting loops around each of which the voltages should sum to zero. The adjustment of such a network is described in Section 1.4.3.

## 6.2 Resistivity profiling

Resistivity traverses are used to

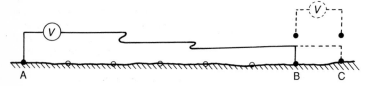

**Fig. 6.2** Establishing a new SP base (B). The value at B relative to A is measured directly and also indirectly via measurements of the voltage at C from both bases.

detect lateral changes. Array parameters are kept constant and the depth of penetration therefore varies only with changes in subsurface layering.

## 6.2.1 Targets

The ideal traverse target is a steeply-dipping contact between two rock types of very different resistivity, concealed under thin and relatively uniform overburden. Such targets do exist, but more often the geological changes of interest cause only small changes in apparent resistivity which must be distinguished from a background due to other geological sources. Targets such as gravel lenses in clays, ice lenses in Arctic tundra and caves in limestone are all much more resistive than their surroundings but are rather small and so may be difficult to detect. Electromagnetic methods (Chapter 8) are usually more effective than d.c. when small, very good conductors such as sulphide ore bodies are being sought.

Depth information can be obtained from a profile if only two layers, of known and constant resistivity, are involved since each value of apparent resistivity can then be converted into a depth using a two-layer type-curve (Fig. 6.3). The estimates should be checked at regular intervals against the results of expanding-array soundings.

## 6.2.2 Choosing arrays

Cables and electrodes are moved long distances on resistivity traverses and the arrays preferred are those for which such moves can be made as simple as possible. The gradient array, with only two moving electrodes, has much to recommend it but the size of the area that can be covered is dictated by the power available and is small unless current is supplied by motor generators.

With the Wenner array, all four electrodes are moved but since the distances between them are all the same, mistakes are unlikely. Entire traverses of cheap metal electrodes can be laid out in advance. Provided that d.c. or very low frequency a.c. is used, so that induction is not a problem, traversing can be speeded up by cutting cables to the desired lengths and binding them together.

The dipole–dipole array is mainly used in IP work (Chapter 7), where induction effects must be avoided at all costs. It uses four moving electrodes and the observed voltages tend to be rather small.

## 6.2.3 Traverse field-notes

Array parameters remain the same along a traverse, hence array type and spacing, and very often current settings and voltage ranges, can be listed on page headers. Only station numbers, $V/I$ readings and remarks need be recorded at individual stations. Current and voltage ranges should be noted if they vary, since they indicate reading reliability.

Comments should be made on changes in soil type, vegetation or

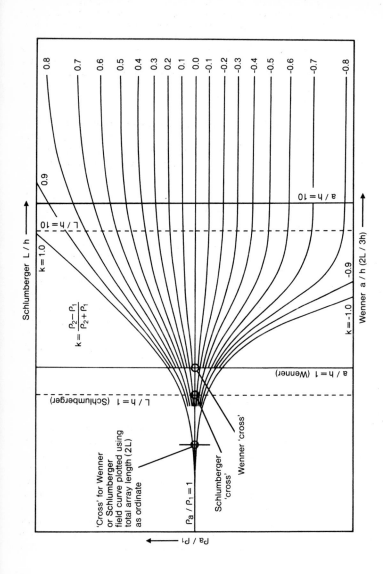

**Fig. 6.3** Two-layer apparent-resistivity type curves for the Wenner array, plotted on log–log paper. When matched to a field curve obtained over a two-layer Earth, the line $a/h = 1$ indicates the depth to the interface and the line $\rho_a/\rho_1 = 1$ indicates the resistivity of the upper layer. The match gives a value of $k$ for the field curve, from which the value of $\rho_2$, the resistivity of the lower layer, can be calculated. The curves can also be used approximately for the Schlumberger array, with the depth to the interface being given by the line $L/h = 1$. Schlumberger and Wenner field curves can both be plotted using total array length as the measure of expansion, in which case the leftmost 'cross' would be used to estimate depth, accurately for the Wenner with this set of curves, approximately for the Schlumberger.

topography and on cultivated or populated areas where non-geological effects may be encountered. Inevitably, these notes will be made by the instrument operator, but will of course relate to readings made with any of the electrodes near the point in question. Since each note which is specific to a certain field point will tend to describe it in relation to the generality of stations, a regional description should be included.

*6.2.4 Displaying traverse data*

The results of resistivity traverses are logically displayed as profiles, which preserve all the features of the original data. Profiles of resistivity and topography can be presented together, along with abbreviated versions of the field notes. Data over an area can be shown by plotting profiles on a base map, but there will usually not then be room for much annotation.

The strike-directions of resistive or conductive features may be shown more clearly by contours than by stacked profiles. Traverse lines and data-point locations should always

be indicated on contour maps since maps of the same area produced using arrays extended in different directions can be very different.

**6.3 Resistivity depth-sounding**

Resistivity depth-soundings investigate layering, using arrays in which the distances between some or all of the electrodes are systematically increased. Apparent resistivities are plotted against expansion, on log–log paper. Although techniques have been proposed for dip interpretation, depth-sounding works well only where the interfaces are roughly horizontal.

*6.3.1 Selection of arrays*

Since depth-sounding involves expansion about a centre point, the instruments generally stay in one place. Portability is therefore less important than it is in profiling but, even so, the Schlumberger array, in which only two electrodes are moved, is often preferred for speed and convenience. Much interpretational literature and many type curves are available.

The Wenner array is also very popular. Interpretational material is plentiful and in fact the differences between Wenner and Schlumberger curves are usually below the level of observational noise (see Fig. 6.3). The effects of near-surface inhomogeneity may be reduced using offsetting techniques.

The dipole–dipole array is virtually never used for ordinary sounding work, although this is possible in principle. Dipole–dipole depth pseudo-sections, much used in IP surveys, are discussed in Section 7.4.2.

### 6.3.2 Using the Schlumberger array

Site selection, extremely important in all sounding work, is particularly critical with the Schlumberger array, which is very sensitive to conditions around the closely-spaced inner electrodes. A location where the upper layer is very inhomogeneous is unsuitable for an array centre, and the offset Wenner may be preferred in land-fill areas.

Directions of expansion are often constrained by topography; there may be only one direction in which electrodes can be taken a sufficient distance in a straight line. If there is a choice, an array should be expanded parallel to the probable strike direction, so that the effect of non-horizontal bedding is minimized. It is generally desirable for there to be a second, orthogonal expansion to check for directional effects, even if

only a very limited line length can be obtained.

The outer electrodes of a Schlumberger array are usually moved in steps which are approximately or accurately logarithmic. The half-spacing sequence 1.5, 2, 3, 5, 7, 10, 15 ... has been very popular in the past, but interpretational computer programs may require exact logarithmic spacing. At six readings to the decade the sequence becomes 1.47, 2.15, 3.16, 4.64, 6.81, 10, 14.7 ...; at five to the decade it is 1.58, 2.51, 3.98, 6.31, 10, 15.8 .... Curves drawn through readings at other spacings can be re-sampled but there are obvious advantages in being able to use the field results directly.

Apparent resistivities are usually calculated from the approximate equation of Fig. 5.3c, which strictly applies only when the inner electrodes form an ideal dipole of negligible length. Although a more accurate apparent resistivity can be obtained using the precise form, the resulting curve is not necessarily more interpretable, since all the type curves are based on dipole models.

Eventually the voltage drop will become too small to be accurately measured and the inner electrodes must then be moved further apart. The sounding curve will thus consist of a number of separate segments (Fig. 6.4). Even if the ground actually is divided into layers which are perfectly internally homogeneous, the segments will not join smoothly because the approximations made in using the dipole equation are different for different $l/L$ ratios. The effect

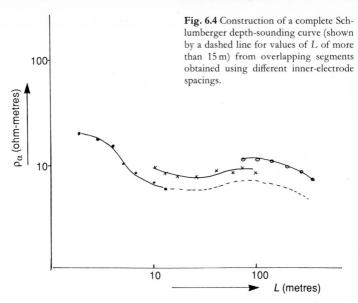

**Fig. 6.4** Construction of a complete Schlumberger depth-sounding curve (shown by a dashed line for values of $L$ of more than 15 m) from overlapping segments obtained using different inner-electrode spacings.

is generally less important than the effect of variation in ground inhomogeneity around the potential electrode positions, and the segments may be linked for interpretation by moving them in their entirety parallel to the resistivity axis so that a continuous curve is formed. To do this, overlap readings must be made; ideally there should be at least three such at each change-over, but two are more usual and one is unfortunately the norm.

### 6.3.3 Offset Wenner depth-sounding

Schlumberger interpretation is complicated by the segmentation of the sounding curve and by the use of an array which only approximates the conditions assumed in calculating type curves. With the Wenner array, near-surface conditions differ at all four electrodes for each reading, giving a rather high noise level. A very much smoother sounding curve can be produced using offsetting techniques.

In offset sounding, five equispaced electrodes are set out, although only four are used for any one reading. Two readings are taken at each expansion (Fig. 6.5a). Averaging the two produces a curve in which local effects are suppressed, while differencing provides estimates of the significance of those effects.

The use of five electrodes might seem to complicate field work, but if expansions are based on the doubling of previous spacings (Fig. 6.5b), very

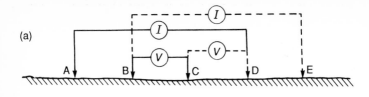

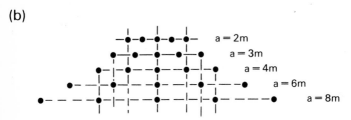

a = 2m

a = 3m

a = 4m

a = 6m

a = 8m

**Fig. 6.5** Offset Wenner sounding. (*a*) Readings are obtained with outer electrodes at A and D and inner electrodes at B and C, and also with outer electrodes at B and E and inner electrodes at C and D. The average of the two readings provides the value to be plotted on the sounding curve and their difference an estimate of near-surface inhomogeneity. (*b*) Expansion system for an offset Wenner survey allowing re-use of electrode positions and hence efficient use of a multicore cable.

quick and efficient operation is possible. This is especially true if special multicore cables are used.

### 6.3.4 Depth-sounding notebooks

In field notebooks, each sounding should be identified by location, orientation and array type. The general environment should be clearly described and any peculiarities, such as the reasons for the

choice of a particular orientation, should be given. Considerable variations in current strengths and voltage levels are likely, and range-switch settings should be recorded for each reading.

Generally, and particularly if a Schlumberger array is used, the operator will be able to see the inner electrodes. For comments on the outer electrodes he must often either rely on second-hand reports or walk the whole line himself. Which he does should depend on the magnitude of the variations being observed and on the experience and reliability of the field crew.

### 6.3.5 Presentation of sounding data

There is usually time whilst distant electrodes are being moved for apparent resistivities to be calculated and plotted. Minor delays are in any case better than returning with unin-

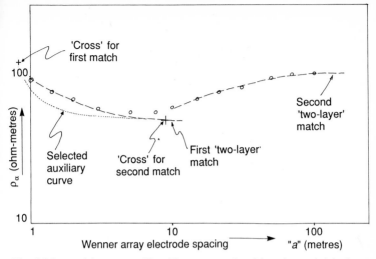

**Fig. 6.6** Sequential curve-matching. The curve produced by a low-resistivity layer between two layers of higher resistivity is interpreted by two applications of the two-layer curves. In matching the later part of the curve, the intersection of the $a/b = 1$ line and the $\rho_\alpha/\rho_1 = 1$ line must lie on a line defined by use of an auxiliary curve.

terpretable results, and field plotting should be routine. All that is needed is a pocket calculator and a supply of log–log paper.

Simple interpretation can be carried out using two-layer type curves (Fig. 6.3) on transparent material. Usually an exact two-layer fit will not be found and a rough interpretation based on segment by segment matching will be the best that can be done (Fig. 6.6). The process is controlled using auxiliary curves to define the allowable pos-

itions of the origin of the two-layer curve being fitted to the later segments of the field curve. Books of three-layer curves are also available, but a full set of four-layer curves would fill a small room.

Step-by-step matching was the main interpretation method used until about 1980. Since that time microcomputer interactive modelling has been possible, even in field base camps, and gives more reliable results.

# 7

# Induced polarization

When current flows in the ground, some parts of the rock mass become electrically polarized. The process is analogous to charging a capacitor or a re-chargeable battery, and both capacitative and electrochemical effects are involved.

If the current suddenly ceases, the polarization cells discharge over a period of several seconds, producing currents, voltages and magnetic fields which can be detected at the surface. Induced polarization (IP) arrays are similar to those used in conventional resistivity surveys. The gradient array (for reconnaissance) and the dipole–dipole array (for detailed work) are especially popular because current and voltage cables can be well separated to avoid induction effects. Current electrodes are normally metal stakes but non-polarizing voltage electrodes must be used to detect the few millivolts of transient signal.

Disseminated sulphide minerals can produce large polarization effects and IP techniques are widely used in exploring for base metals.

## 7.1 Polarization fundamentals

There are two main mechanisms of rock polarization and two main ways in which the effects can be observed. Time and frequency measurements provide equivalent results in theory but there are practical differences.

### 7.1.1 Membrane polarization

Clays and some other platey or fibrous minerals have negatively-charged surfaces which cause *membrane polarization* in rocks with small pore spaces. Positive ions in the pore waters of such rocks concentrate near the pore walls. If an electric field is applied, the positive-ion clouds are distorted and negative ions move into them and are trapped, producing concentration gradients which impede current flow. When the applied field is removed, a reverse current flows to restore the original equilibrium.

### 7.1.2 Electrode polarization

In addition to the static 'contact potentials' between a metallic conductor and an electrolyte, over-voltages are produced whenever currents flow. Despite its name, this *electrode polarization* occurs not merely at artificial electrodes but wherever grains of electronically-conducting minerals are in contact with the ground water.

The degree of polarization is determined by areas of contact and not by the mass of mineral involved, and polarization methods are thus exceptionally well suited to exploration for disseminated 'porphyry' mineralization. Massive sulphide ores usually produce strong anomalies because of a surrounding disseminated 'halo'.

Although, given equivalent areas of active surface, electrode polarization is the stronger mechanism, clays are much more abundant than sulphides and most IP effects are due to membrane processes.

### 7.1.3 The square-wave in chargeable ground

When a steady current is suddenly terminated, the voltage $V_0$ observed between two grounded electrodes drops abruptly to a small value $V_p$, and then declines asymptotically to zero. A similar effect can be observed when current is applied to the ground; the measured voltage rises rapidly at first and then approaches $V_0$ asymptotically (Fig. 7.1). Although in theory $V_0$ is never actually reached, in practice the difference is not detectable after a few seconds.

Chargeability is defined theoretically as the polarization voltage developed across a unit cube energized by a unit current and is thus in some ways analogous to magnetic susceptibility. In terms of the signal shown in Fig. 7.1, chargeability is defined as the ratio of $V_p$ to $V_0$. This

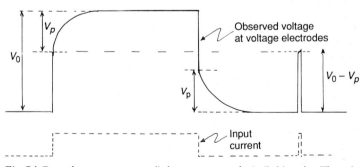

**Fig. 7.1** Ground response to an applied square wave and a 'spike' impulse. The ratio of $V_p$ to $V_0$ is seldom more than a few percent.

*apparent chargeability* of the entire rock mass is a pure number, without units, but for convenience is generally multiplied by a thousand and quoted in millivolts per volt.

The $V_p/V_0$ ratio cannot be measured directly since electromagnetic transients are dominant in the first tenth of a second. Time-domain chargeability is therefore defined in terms of the decay voltage at some later time and the link with the theoretical definition becomes rather tenuous. Not only do different instruments use different delays, but it was originally essential and is still quite common to measure an area under the decay curve using integrating circuitry, rather than an instantaneous voltage. Where integration is used, results depend on the length of the integration period as well as the delay and are quoted in milliseconds.

### 7.1.4 Frequency effects

If, in a resistivity survey using a square-wave input, the applied field is cut off before the voltage $V$ reaches $V_0$, a lower apparent resistivity (equal to $V/I$ multiplied by the appropriate geometrical factor) will be calculated. The frequency effect is defined as the difference between the 'high frequency' and 'd.c.' resistivities, divided by the high-frequency value. This is multiplied by 100 to give an easily handled whole number, the *Percent Frequency Effect* (PFE).

The theoretical frequency effect

calculated for a current which flows only instantaneously to produce a voltage $V_0 - V_p$ (Fig. 7.1), is related to chargeability by the equation:

$$M = \frac{(PFE)}{100 + (PFE)} \tag{7.1}$$

Because of electromagnetic transients, this 'instantaneous' resistivity cannot be measured and the practical PFE depends on the frequencies used. To cancel telluric and SP noise, even the 'd.c.' measurement is taken with current alternated at about 0.2 Hz.

### 7.1.5 Metal factors

The PFE can be divided by the d.c. resistivity to yield a quantity which, multiplied by 1000, 2000 or $2000\pi$, produces a convenient number, known as the *metal factor*.

Metal factors emphasize rock volumes that are both polarizable and conductive and which may therefore be assumed to have a significant sulphide (or graphite) content. Although this may be useful when dealing with massive sulphides, low resistivity is irrelevant or actually misleading in exploration for disseminated deposits. As usual when factors which should be considered separately are combined, the result is confusion, not clarification.

### 7.1.6 Phase and spectral measurements

The asymmetry of the voltage curve

in Fig. 7.1 implies a phase shift between applied field and voltage. This shift can be measured directly to give polarization anomalies in milliradians. If a number of different frequencies are used, a polarization spectrum can be plotted. Most attempts to distinguish between different types of IP source are now based on analysis of these spectral curves.

### 7.1.7 Comparison of time- and frequency-domain methods

The relative merits of time- and frequency-domain IP have long been argued, especially by the manufacturers of competing instruments. It is generally conceded that PFEs are more vulnerable to electromagnetic interference than are time-domain chargeabilities. Although induction effects observed at high frequencies can be used to calculate correction factors for low-frequency data, the extra readings required demand more sophisticated instruments and slower fieldwork.

Time-domain surveys are essentially multi-frequency and the information derivable from decay curve shapes would require measurements at several different frequencies in frequency-domain or phase work. However, time-domain currents and voltages must be large, and frequency surveys may be preferred as being safer and using more portable instruments. The final choice of method usually depends on personal preference and instrument availability.

## 7.2 Time-domain IP surveys

Since time-domain IP surveys measure the small transient voltages which exist after the primary field is removed, large primary voltages are necessary, and even so great care must be taken to minimize telluric effects and voltage-electrode polarizations.

### 7.2.1 Time-domain transmitters

A time-domain transmitter consists of a power source, which may be a large motor generator or a re-chargeable battery pack, connected to an output controller. Voltage levels are usually selectable within a range from 100 to 500 V. Current levels, which may be controlled automatically, must be recorded since apparent resistivities are calculated as well as IPs.

Current direction is reversed after each reading to minimize the effects of natural voltages, and cycle times can generally be varied from 2 to 16 seconds. One second each for energization and reading is not generally sufficient for reliable results, while cycles more than 8 seconds long unreasonably prolong the survey work.

### 7.2.2 Time-domain receivers

A time-domain receiver measures primary voltage and one or more decay voltages or integrations. It may also be possible to record the

SP, so that chargeability, resistivity and SP data can be gathered together. Early Newmont-type receivers integrated in the ranges shown in Fig. 7.2. The $L$ parameter was intended to characterize the decay-curve shape and was usually recorded only where $M$ varied significantly.

With Newmont receivers the SP was first balanced out manually and the primary voltage was then 'normalized' by adjusting an amplifier control until a galvanometer needle swung between precisely defined limits. This automatically ratioed $V_p$ to $V_0$ for the $M$ values recorded by a second needle. Other early receivers worked on similar principles. Analogue displays have considerable advantages for experienced operators: the rates of needle movement provide some 'feel' for the shape of the decay curve and electromagnetic transients, at their largest in the period prior to integration, can often be noted.

The cycle-time set on a receiver must correspond to that programmed into the transmitter so that the receiver can 'lock-on' to the transmission without use of a reference cable. Changing the cycle time can produce quite large differences in apparent chargeability, even for similar delay times; 'binary' cycles of 4, 8 or 16 seconds are now generally favoured but even so, chargeabilities recorded by different instruments are only vaguely related.

With purely digital instruments, the diagnostic information and 'feel,' provided by a moving needle is lost.

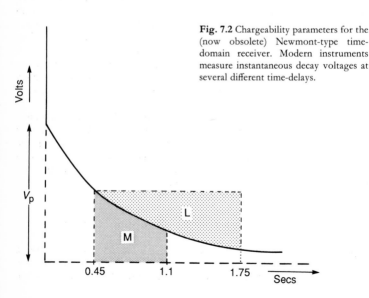

**Fig. 7.2** Chargeability parameters for the (now obsolete) Newmont-type time-domain receiver. Modern instruments measure instantaneous decay voltages at several different time-delays.

Greater reliance must then be placed on a statistical approach to the readings and a sufficient number of cycles must be observed for such an approach to be valid. Digital systems allow more parameters to be recorded and very short integration periods, equivalent to instantaneous readings, to be used. Natural SPs are now 'backed-off' or 'bucked-out' automatically rather than manually. Inevitably, memories have been added to store data and reduce note-taking.

### 7.2.3 Decay-curve analysis

The $L/M$ ratio was intended to provide an indication of the composition of a chargeable body. In practice it does little more than reinforce the correlation of high chargeabilities with sulphides and weaker ones with clays, and does not distinguish between sulphides and graphite.

With several readings on the decay curve, more detailed analysis can be attempted. A method suggested for use with Huntec receivers assumes that decay curves are combinations of two exponential decays, corresponding to electrode and membrane polarizations, which can be isolated mathematically. This is far too drastic a simplification, and the separation of two exponential components using a limited number of readings is in any case virtually impossible in the presence of small amounts of noise.

Studies of decay-curve shapes continue, and chargeabilities should be recorded at as many decay times as are conveniently possible in areas of interesting anomaly. In non-anomalous areas a single value generally suffices.

## 7.3 Frequency-domain surveys

Quite small currents and voltages can be used for resistivity measurements, and so frequency-domain transmitters can be light and portable. Especial care has to be taken in positioning cables to minimize electromagnetic coupling.

### 7.3.1 Frequency-domain transmitters

Square-waves are commonly used for frequency as well as time-domain work and most modern IP transmitters can be used for either. Measurement of resistivity at two frequencies in separate operations is time-consuming and does not allow precise cancellation of random noise. Noise can be cancelled if simultaneous readings are made using a square-wave (if the receiver can analyze the voltage waveform to extract the high-frequency effects) or a complex waveform made up of two frequencies.

### 7.3.2 Frequency-domain receivers

Sophisticated receivers are needed to analyse waveforms and extract frequency effects from either single- or dual-frequency transmissions. This sophistication is normally not apparent to the operator recording the PFEs from a front panel display.

To measure phase differences, usually over a range of frequencies, a time reference must be maintained between transmitter and receiver. Direct linkage is possible, but a reference cable can be a source of inductive coupling and may also be operationally inconvenient. Modern receivers use crystal clocks which can be synchronized with the transmitter at the start of a day's work and which will drift no more than a fraction of a millisecond in twenty-four hours.

## 7.4 Handling IP data

The methods used to display IP data

vary with the array. Profiles or contour maps are used for gradient arrays, while dipole–dipole data are almost always presented as pseudo-sections.

### 7.4.1 Assessing gradient-array results

Gradient-array profiles may be interpreted by methods analogous to those used for magnetic data, since current flows roughly horizontally in the central area and bodies will be horizontally polarized. Depths can be estimated approximately using the techniques of Section 3.5.2.

### 7.4.2 Presenting dipole–dipole data

Dipole–dipole traverses at a single $n$

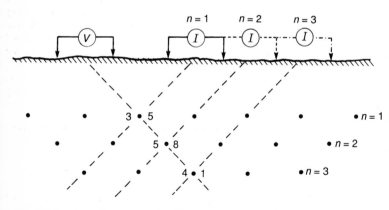

**Fig. 7.3** Pseudo-section construction. The three current–dipole positions correspond to three different multiples of the basic spacing. The measured values (of IP or resistivity) are plotted at the intersections of 45° slope lines from the centres of the current and voltage dipoles. Often, the plotting point doubles as a decimal point for the IP values.

value can be used to construct profiles but multi-spaced results are almost always displayed in the form of pseudo-sections (Fig. 7.3). Interpretation of pseudo-sections in terms of depth distribution is never straightforward and requires considerable experience and familiarity with model studies. In particular, the direction of apparent dip is not necessarily the direction of dip of the chargeable body. The very common 'pant's leg' anomaly is usually produced by a small near-surface body with little extent in depth, since every measurement made with either the current or the voltage dipole near the body will record high chargeability (Fig. 7.4).

Pseudo-sections are nearly always plotted in the stylized form of Fig. 7.3, even in very rugged terrain. Cross-sectional plots drawn against actual topographic profiles have been objected to as suggesting a much closer relationship between pseudo-sections and the real distribution of resistivity in the sub-surface than in fact exists. This is true, but rough topography does influence dipole–dipole results and it is better that it be shown than ignored.

Multi-spaced dipole–dipole data can be converted to single values which can then be contoured or profiled (Fig. 7.5). A great deal of information is lost and the process is only occasionally useful.

### 7.4.3 Negative IPs and masking

Negative IP effects are sometimes recorded. These can be caused by power or telephone cables but signal contribution sections (Fig. 5.5) show that they can also arise from lateral inhomogeneities. Layering can also produce negative values or mask deeper sources, most readily if both the surface and target layers are more conductive than the rocks in between. In these circumstances, the penetration achieved may be very small and the total length of the dipole–dipole arrays used may need to be ten or more times the desired exploration depth.

Interactions between conduction and charge in the Earth are very

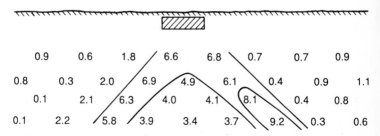

**Fig. 7.4** 'Pant's leg' pseudo-section, showing IP anomaly due to a small, shallow body.

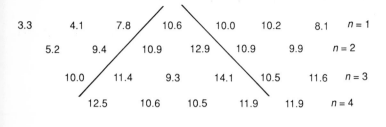

$$\text{Composite value} = \begin{bmatrix} 10.6 \\ + \\ (10.9 + 12.9)/2 \\ + \\ (11.4 + 9.3 + 14.1)/3 \\ + \\ (12.5 + 10.6 + 10.5 + 11.9)/4 \end{bmatrix} \div 4 = \begin{bmatrix} 10.6 \\ + \\ 11.9 \\ + \\ 11.6 \\ + \\ 11.4 \end{bmatrix} \div 4 = 11.4$$

**Fig. 7.5** Filtering of IP dipole–dipole data to produce contourable 'pseudo-gradient array' values.

complex, and interpreters generally need more reliable resistivity data than they actually get, since the dipole–dipole array performs poorly in defining layering. A small number of Wenner or Schlumberger expansions, carried out specifically to map resistivity, may prove invaluable. Also, it is vital that any changes in surface conditions which might correlate with changes in surface conductivity are recorded in the field. The detectability of ore beneath a swamp and beneath bare rock ridges on either side will be quite different.

The relationship between polarization and current is not entirely linear. This not only limits the extent to which time, frequency and phase measurements can be interrelated, but can also affect comparisons between different surveys of the same type. The effects generally do not exceed a few percent, but provide yet another reason for the very qualitative nature of most IP interpretation.

# Electromagnetic methods

The voltages induced by varying electric currents are noise in d.c. surveys but can be used directly to locate good conductors. Wave effects are important only in very low frequency (VLF) work (Chapter 9), and electromagnetic (e.m.) techniques can otherwise be understood in terms of currents in conductors and magnetic fields in space.

## 8.1 Induction principles

Circular, concentric lines of magnetic field surround a current-carrying wire. Bent into a small loop, the wire becomes a magnetic dipole. The field can be varied by alternating the current (continuous wave or c.w. methods) or by terminating it (transient methods or TEM).

### 8.1.1 Induction

In a varying magnetic field, voltages (electromotive forces or emf's) are induced at right-angles to the field, and currents will flow in conductors forming parts of closed circuits. Geological conductors are very complex and for theoretical analyses the 'eddy currents' which flow in them must be approximated using greatly simplified models.

Emf's are determined by the rate of change of current in the inducing circuit and by the value of a parameter known as the *mutual inductance*. Mutual inductances are large, and conductors are said to be well coupled, if there are long adjacent current paths, if the magnetic fields are at right-angles to directions of easy current flow and also if magnetic materials are present to enhance field strengths.

Current change in a circuit induces emf's opposing the change. A circuit such as a tightly wound coil which strongly resists current changes is said to present a high impedance to alternating current and to have a large self-inductance.

### 8.1.2 System descriptions

In both continuous-wave and TEM

surveys, sources (usually) and receivers (always) are wire loops or coils. The form of anomaly depends on system geometry as well as on the nature of the ground conductor.

Coils are described as horizontal or vertical according to the plane in which the windings lie; thus, the axis of a 'horizontal' coil is vertical. Systems are also characterized by whether receiver and transmitter coils are co-planar, co-axial or orthogonal, and by whether the coupling between them is at a maximum, a minimum or variable. *Orthogonal coils* are minimum-coupled and the primary field is not detected. *Co-planar* and *co-axial coils* are maximum-coupled, since the primary flux from the transmitter lies along the axis of the receiver coil (Fig. 8.1); such systems are scarcely affected by slight coil misalignments but, because a strong in-phase field is detected even in the absence of a conductor, they are very sensitive to changes in coil separation.

Dip-angle systems, in which the receiver coil rotates to determine the dip of the resultant field, were once very popular. They are now generally limited to VLF work and to the Crone 'Shoot-back' system in which topographic effects can be eliminated by interchanging receiver and transmitter.

*8.1.3 Phase*

Geological bodies are neither perfect conductors, in which current would flow in phase with the inducing field,

nor pure inductances, but have both inductance and resistance. The currents induced, and the associated 'secondary' magnetic fields, will differ in phase from the primary field; the resultant at the receiver will have components in-phase and 90° out of phase with the primary (Fig. 8.1d).

In-phase signals are sometimes also termed 'real', the other component being described as imaginary, quadrature or simply out-of-phase.

Since magnetic fields propagate at the speed of light and not instantaneously, phase will change with distance from the transmitter. The shifts are trivial, and are ignored, in all normal surveys.

## 8.2 Continuous-wave systems

In ground surveys the use of horizontal co-planar coils is almost standard and there is usually a shielded cable carrying a phase-reference signal from transmitter to receiver. The sight of the two operators, each encumbered by bulky apparatus and linked by a sort of umbilical cord, struggling across rough ground and through thick scrub has provided comic relief on many surveys. Some instruments allow the reference cable to be used for voice communication; this is an excellent idea, but it is just as well that memory units have not (yet) been developed to record the conversations.

The Swedish term *Slingram* is often applied to horizontal-loop systems but without any general

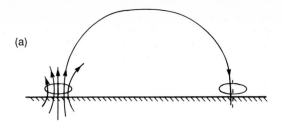

(a)

(b)

(c)

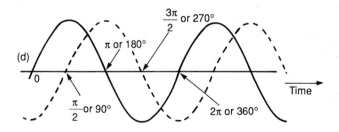

(d)

**Fig. 8.1** Coil-systems. (*a*) Horizontal, coplanar (maximum-coupled). (*b*) Vertical, coaxial (maximum-coupled). (*c*) Orthogonal (minimum-coupled). (*d*) Quadrature relationship. The phase difference is 90° or $\pi/2$ radians.

agreement as to whether it is the use of two mobile coils, or their coplanar and horizontal orientations, or the use of a reference cable which makes the term applicable.

### 8.2.1 Response functions

The electromagnetic response of a geological body is proportional to its mutual inductances with the transmitter and receiver circuits, and inversely proportional to its self-inductance, which limits eddy current flow. Because horizontal-loop amplitude anomalies are expressed as percentages of primary field, response is effectively also inversely proportional to the mutual inductance between transmitter and receiver. These four parameters can be combined in a single 'coupling factor'.

Anomalies also depend on conductivity and frequency, as combined in a 'response parameter' involving frequency, self-inductance (always closely related to the linear dimensions of the body) and resistance. Response curves (Fig. 8.2) may be viewed as showing either how response will vary over targets of different resistivity using a fixed-frequency system or over one target as frequency is varied. At high frequencies the distinction between good and merely moderate conductors tends to disappear. The quadrature field is very small if frequencies and conductivities, and hence eddy-currents, are small, and also at high frequencies and high conductivities.

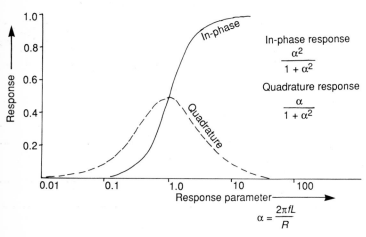

In-phase response

$$\frac{\alpha^2}{1 + \alpha^2}$$

Quadrature response

$$\frac{\alpha}{1 + \alpha^2}$$

$$\alpha = \frac{2\pi fL}{R}$$

**Fig. 8.2** Response of a horizontal-loop e.m. system to a vertical-loop target, plotted as a function of the parameter $2\pi fL/R$. Horizontal scale is logarithmic. L is the loop self-inductance, R its resistance and f is the frequency.

## 8.2.2 Choice of survey parameters

Induction implies loss of energy from the varying field, and conduction in the overburden therefore reduces penetration. At high frequencies, all overburdens appear conductive and penetration, characterized by the skin depth (Fig. 8.3), is small. Single-frequency prospecting systems therefore operate below 1000 Hz. Multi-frequency systems are now very generally used, since they allow conductivity to be estimated, but even these are normally operated entirely below 5000 Hz. Poor-quality conductors may produce anomalies only at the highest frequency or not at all (see Fig. 9.10).

At low frequencies and reasonable values of surface-layer conductivity (i.e. where skin-depth is not a major consideration), the depth of investigation is largely determined by coil spacing. An increase in spacing attenuates the primary signal more than it does the signal from a deep conductor, because of the operation of the inverse cube law (Fig. 8.4). The coupling factor thus increases with coil separation but, inevitably, system resolution deteriorates because of the larger volume of rock being sampled. Estimates of depth penetration are even more problematic in e.m. than in d.c. surveys. A common rule of thumb estimates the penetration as equal to half the separation, but this is unduly optimistic in most circumstances.

Coupling also depends on orientation. Lines should be laid out across the expected strike of the targets. Oblique intersections produce poorly defined anomalies which may be difficult to interpret.

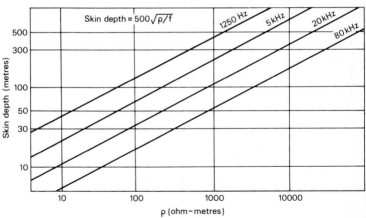

**Fig. 8.3** Variation of skin-depth, $d$, in which the amplitude of an electromagnetic signal is reduced to $1/e$ ($=0.368$) of its surface value, with frequency and resistivity. If $d$ is in metres and $\rho$ in ohm-metres, $d = 504\sqrt{\rho/f}$.

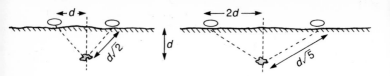

**Fig. 8.4** Spacing and penetration. When the two coils are moved apart, the fractional change in distance between them is greater than that between either one of them and the conductor at depth, so that increases in spacing increase the anomalous field as a percentage of the primary. In the example shown, doubling the coil spacing increases the coil-to-target distance by only about 60%.

### 8.2.3 Survey procedures

Because signals are ratioed to primary field strength, the '100%' level must be set on the instrument at the start of each day by reading at the standard survey spacing on ground which is level and believed to be non-anomalous. This check should be carried out even with instruments that have 'click-stop' settings for the allowable separations. A check for 'phase-mixing', i.e. for leakage of a percentage of the primary signal into the quadrature channel, should also be made. Handbooks should describe both how to make this check and how to make any necessary adjustments.

Most surveys are carried out using at least two frequencies, and both receiver and transmitter must be tuned to the same frequency for sensible readings to be obtained. Receivers can be seriously damaged if a transmitter tuned to their frequency is operated within a few feet of them.

Figure 8.5 illustrates the anomaly observed when the horizontal-loop

system is taken across the strike of a steeply dipping conducting 'sheet'. Since current can circulate only in the plane of the sheet, the secondary field immediately above it is horizontal and no anomaly will be detected there by a horizontal receiving coil; similarly, there will be zero anomaly when the transmitter is vertically above the sheet because no significant eddy currents will be induced. The greatest (negative) secondary field values will be observed when the conductor lies half-way between the two coils.

Readings obtained with mobile transmitter and receiver coils are plotted on maps at points mid-way between the two. This is realistic because in most cases where relative coil-orientations are fixed, the anomaly profiles over symmetrical bodies are symmetrical and are not affected by interchanging receiver and transmitter. Even where this is not completely true, recording the mid-point is less likely to lead to confusion than using transmitter or receiver positions.

In all electromagnetic work, care

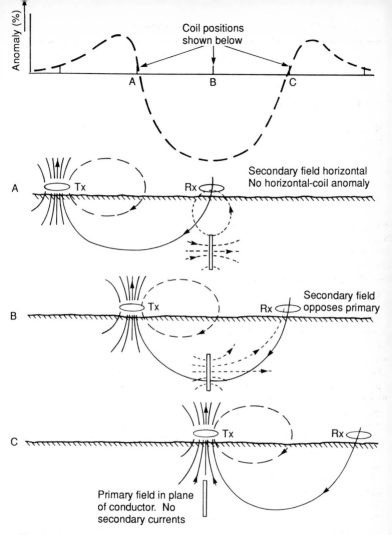

**Fig. 8.5** Anomaly obtained with a horizontal-loop system across a near-vertical conducting sheet. Over a dipping sheet the area between the side-lobe and the 100% line would be greater on the down-dip side. Note that the width of the anomaly is largely determined by the coil spacing and not by the conductor width or depth. Tx = Transmitter; Rx = Receiver.

113

must be taken to record any environmental variations which might affect the results. Cultural features such as roads, along which buried artificial conductors often run, must be noted, as well as obvious actual conductors. Power and telephone lines cause special problems, since they broadcast noise which, although at a different frequency, is often strong enough to pass through the filters.

Ground conditions should also be noted, since variations in overburden conductivity can drastically modify the shape of an anomaly as well as the penetration of the signal. In many hot, dry countries, such as Australia, salts in the overburden produce surface conductivities so high that c.w. methods are ineffective and have been superseded by TEM.

## 8.2.4 Effects of separation

Spurious in-phase anomalies can be introduced by changes in coupling between source and receiver.

The field a distance $r$ from a coil (magnetic dipole source) can be described in terms of radial and tangential components $F(r)$ and $F(t)$ (Fig. 8.6):

$$F(r) = 2A \sin \theta / r^3$$
$$F(t) = A \cos \theta / r^3$$

The amplitude factor $A$ depends on coil dimensions and current strength.

If the coils are truly co-planar, $F(r)$ is zero (because $\theta$ is zero) and the measured field $F$ is equal to $F(t)$. The consequence of a fractional change $x$ in separation is then that:

$$F = F_0 / (1 + x)^3$$

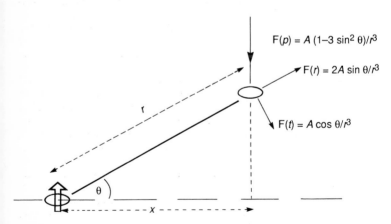

**Fig. 8.6** Field components due to a current-carrying loop acting as a magnetic-dipole source. $F(r)$ and $F(t)$ are radial and tangential components respectively. $F(p)$, obtained by adding the vertical components of both, is the primary field as defined for horizontal-loop slope corrections.

where $F_0$ is the field strength at the intended spacing. If $x$ is small, an approximate expression can be used:

$$F = F_0 (1 - 3x)$$

Thus, for small errors, the percent error in in-phase component is three times the percent error in distance. Since real anomalies of only a few percent can be important, separations must be carefully measured.

### 8.2.5 Surveys in slopes

On sloping ground, the distance between survey pegs may either be measured along slope or the line may be 'secant-chained' so that horizontal distances are constant.

If along-slope distances are used in reasonably gentle terrain, coil separations should be constant. However, it is difficult to keep coils coplanar without a clear line of sight and simpler to hold them horizontal. The field component $F(r)$ along the receiver axis then equals the coplanar field multiplied by $(1 - 3 \sin^2\theta)$. The corresponding correction factor is always greater than 1 (coils really are 'maximum-coupled' when coplanar) and becomes infinite when the slope is 35° and the primary field is horizontal (Fig. 8.7).

If lines are secant-chained, the along-slope distance is proportional to the secant (1/cosine) of the slope angle. For coplanar (tilted) coils the ratio of the 'normal' to the 'slope' field is therefore $\cos^3\theta$. The correction must be divided by $(1 - 3 \sin^2\theta)$ if the coils are held horizontal (Fig. 8.7).

Separations in rugged terrain can differ from their nominal values because slope distances have been measured between points which are more closely spaced than the coil separation in use (Fig. 8.8). Accurate surveying is essential in such areas, and field crews may need to carry listings of the coil tilts required at each station. Instruments which incorporate tilt meters and communication circuits are virtually essential and even so errors are depressingly common.

### 8.2.6 Applying the corrections

For any coupling error, whether caused by distance or tilt, the field $F$ which would be observed with no conductors present can be calculated as a percentage of the maximum-coupled field $F_0$. If $F$ is calculated to be 92% of $F_0$, this can be converted to 100% either by adding 8% or by multiplying by 100/92. If the reading obtained actually is 92%, these two operations will produce identical results. If, however, there is an additional anomalous field, adding 8% will correct only the primary field but multiplication will apply a correction to the anomaly also. Neither procedure is actually 'right', but the deeper the conductor (in relation to the coil spacing), the less the effect of a distance error on the secondary field (see Fig. 8.9). Since to be detected at all the conductor is likely to be quite near the surface, correction by multiplication is generally more satisfactory, and factors

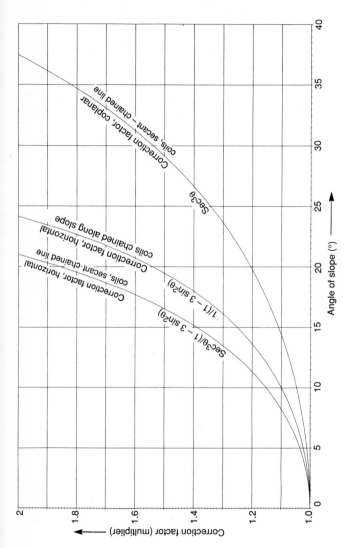

**Fig. 8.7** Slope correction factors for a two-coil system calibrated for use in horizontal, co-planar mode. Readings should be multiplied by the appropriate factors.

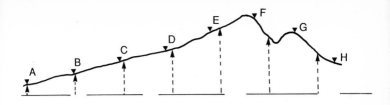

**Fig. 8.8** Secant-chaining and slope chaining. Downward arrows show locations of along-slope stations and upward arrows locations of corresponding secant-chained stations. In the region between points F and G, where topographic 'wavelength' is less than peg spacing, the sum of the slope distances measured in short segments will be greater than the actual coil separation.

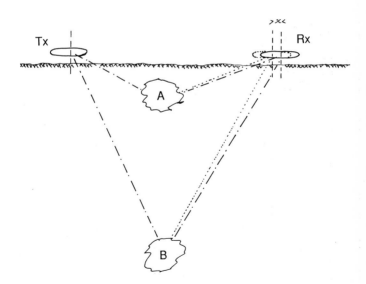

**Fig. 8.9** Effect of coupling errors on secondary fields. The distance error x (which could be due to secant-chaining in rugged terrain) will have a much greater effect on an anomaly produced by a body in position A than by one in position B, because of the percentage change in distance.

**Fig. 8.10** Geonics EM-31 non-contacting ground-resistivity unit. (Photo, courtesy Geonics.)

derived from Fig. 8.7 should be used in this way.

Coupling errors cause fewer problems if only quadrature fields are observed, since these are anomalous by definition (although, as Fig. 8.2 shows, they may be small for very good as well as poor conductors). Approximate coupling-corrections could be made using the in-phase multipliers but there is little point in doing this in the field. The detailed problems caused by changes in coupling between a transmitter, a receiver and a third conductor can, thankfully, be left to the interpreter, provided the field notes describe the system configurations precisely.

## 8.3 Other e.m. techniques

E.m. surveys can be carried out in a variety of other ways, using different system geometries, long-wire sources instead of coils and pulsed signals instead of continuous waves. These can only be considered very briefly.

### 8.3.1 Ground-conductivity measurement

Vertical co-axial coil systems are sensitive to the effects of flat-lying conductors and can be used to estimate ground conductivity without actual ground contact. The Geonics EM-31 is an example of such an instrument which can be used, at some risk to life and limb in some environments, by one man (Fig. 8.10). The apparent resistivity estimates can be completely invalidated by very good, man-made conductors such as wire fences and metal sheeting.

### 8.3.2 Fixed source methods

The fields produced by straight, current-carrying wires can be calculated by repeated applications of the *Biot–Savart Law* (Fig. 8.11). For four wires forming a rectangular loop the relationship illustrated in Fig. 8.12 can be deduced. If the measurement point is outside the loop, radius vectors which do not cut any side of the loop must be given negative signs.

The anomaly shown in Fig. 8.5 is symmetrical because the receiver and transmitter coil in turn move over the body. If the source, whether a coil or a straight wire, were to be fixed, there would be a zero only with the receiver coil vertically above the body and the anomaly would be anti-symmetric (Fig. 8.13, see p. 120). Such anomalies are often observed as dip-angles, ratioing vertical and horizontal fields.

Fixed extended sources are used in *Turam* (Swedish: 'two coils') methods, in which two receiving coils are separated by a distance of, usually, 10 m. The fields are compared in phase and, by calculating a 'reduced ratio' equal to the actual ratio of the signal amplitudes divided by the 'normal' ratio, in amplitude. Any phase difference is anomalous and there is no need for a reference

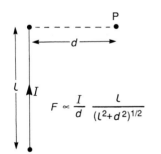
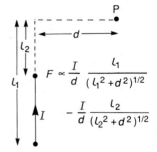

**Fig. 8.11** The Biot–Savart law. Any long-wire system can be regarded as made up of elements of the type shown in (*a*). Two such elements, with current flowing in opposite directions, can be used to obtain an expression for the field in cases such as that shown in (*b*), where P is beyond the end of the wire.

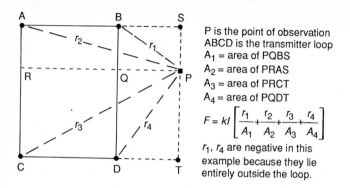

P is the point of observation
ABCD is the transmitter loop
$A_1$ = area of PQBS
$A_2$ = area of PRAS
$A_3$ = area of PRCT
$A_4$ = area of PQDT

$$F = kI \left[ \frac{r_1}{A_1} + \frac{r_2}{A_2} + \frac{r_3}{A_3} + \frac{r_4}{A_4} \right]$$

$r_1$, $r_4$ are negative in this example because they lie entirely outside the loop.

**Fig. 8.12** Primary field due to a fixed, rectangular, conducting loop carrying a current $I$. For $I$ in amps, distances in metres and $F$ in Weber · metre$^{-2}$, $k = 10^{-7}$.

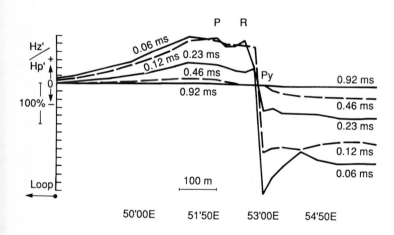

**Fig. 8.13** Fixed-loop UTEM vertical component anomaly at Que River, Tasmania. Reading interval 25 m. The important feature of this profile is the detection of the weak anomaly at P, which indicates mineralization deeper than the barren pyrite which produces the large anomaly at R. (Reproduced courtesy Australian Society of Exploration Geophysicists.)

120

cable to bring phase information from the transmitter. Absolute phases and ratios relative to a single base can be calculated provided that successive readings are taken with the trailing coil placed in the position vacated by the leading coil. Turam is now little used, but large fixed sources are common in TEM work.

### 8.3.3 Pulse-transient methods

TEM systems provide multi-frequency data by sampling transients at a number of different times after the transmitter current is cut off. Since measurements are made when no primary field is present, the transmitter loop, which may have sides of 100 m or more, can be used to receive the eddy-current signal. Alternatively, the absence of primary field during measurement allows the receiver to be positioned within this loop, a technique which can be used in c.w. work only with very large loops because of the strong coupling to the primary field. A system which does not have to be portable and in which the primary field cannot 'leak' into secondary field readings can use very high powers, and TEM systems are popular in areas where overburden conductivities are high and penetration is skin-depth limited.

Conventional arrangements using small transmitter and receiver coils are also common and most commercial TEM systems can be used with several different coil configurations. They differ in portability and, in detail, in sampling programs.

The SIROTEM may be taken as typical; it produces a square-wave current with equal on and off times in the range from 23 to 185 msec. The voltage in the receiver coil can be recorded at 32 time-instants during eddy-current decay. Signals can be averaged over up to 4096 cycles, the results being recorded either on paper or on magnetic tape.

The transmitter current in the UTEM systems originally developed at the University of Toronto has a precisely triangular waveform, with a fundamental frequency in the range from 25 to 100 Hz. In the absence of ground conductivity, the received signal, proportional to the time-derivative of the magnetic field, is a square-wave. Deviations from this waveform in the vertical magnetic and horizontal electric fields produced by a large rectangular loop are observed by sampling at eight time-intervals.

TEM data are usually presented by 'stacking' the profiles for the individual 'channels' (i.e. for each sampling delay time). The results at short delay times are dominated by eddy-currents in large volume, relatively poor conductors; at longer delays these currents have died away and only those in the very good conductors remain (Fig. 8.13).

### 8.3.4 E.m. and IP

TEM surveys have a superficial resemblance to time-domain IP. The most obvious difference between the methods is that in IP the currents are introduced directly into the ground and not induced by magnetic fields. However, at least one IP method uses induction, and a more fundamental difference lies in the time scales.

Time-domain IP systems usually sample after delays of between 100 msec and 2 sec, and so avoid most e.m. effects. There is a small region of overlap, from about 100 to 200 msec, between TEM and IP systems and some frequency-domain or phase IP units are designed to work over the whole range of frequencies from virtual d.c. to tens of kilohertz to obtain conductivity spectra. However, it is quite possible to regard the e.m. and IP phenomena as completely separate and to avoid working in regions, either of frequency or time delay, where both are significant.

# 9

# VLF methods

Geophysical instruments have been developed to take advantage of military, high-power communications transmissions in the 15–25-kHz band. Termed *very low frequency* (VLF) by radio engineers, these waves have frequencies higher than those used in conventional geophysical work, but allow electromagnetic surveys of a kind to be carried out without local transmitters.

**Fig. 9.1** Electromagnetic-wave vectors close to a perfect conductor. The magnetic (H) and electric (E) fields are at right angles to each other and to the direction of propagation, and differ in phase by 90°, so that one is zero when the other is a maximum.

## 9.1 VLF radiation

An electromagnetic wave consists of coupled alternating electrical and magnetic fields, directed at right angles to each other and to the power vector which defines the direction of propagation (Fig. 9.1). Electric field vectors will always align themselves at right angles to perfectly conductive surfaces and a wave can therefore be 'guided' by enclosing conductors. The extent to which this is possible is governed by the relationship between the wavelength of the radiation and the dimensions of the guide.

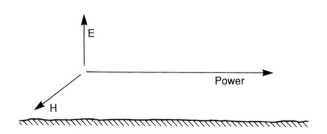

### 9.1.1 VLF transmissions

There are more than a score of stations around the world which transmit VLF signals continuously for military purposes (Fig. 9.2). The waves are ideal for communicating with submarines, since they can be detected tens of metres below the sea surface and propagate very efficiently over long distances in the waveguide formed by the Earth and the ionosphere. The message content is generally superimposed by frequency modulation on a sinusoidal carrier wave, but occasionally the carrier is chopped into dots and dashes resembling Morse code. Making geophysical use of these 'quenched-carrier' signals is extremely difficult. Transmission patterns and servicing schedules vary widely. The makers of VLF instruments are usually aware of the current situation and provide information to their customers on request.

Neither the Earth nor the ionosphere is a perfect conductor, and some VLF energy is lost into space or penetrates the surface. Without this penetration, the waves would have neither military nor geophysical uses. Amplitudes decrease exponentially with depth and the secondary fields produced in subsurface conductors are similarly attenuated on their way to the surface. The relationship between skin depth and ground resistivity for a 20-kHz wave was shown in Fig. 8.3.

### 9.1.2 Detecting VLF fields

A geophysical user of a VLF signal has control over neither the amplitude at the transmitter nor the phase. Readings of a single field component at a single point are therefore meaningless; one component must be selected as a reference with which the strengths and phases of other components can be compared. The obvious choices are the horizontal magnetic or vertical electric fields, since these approximate most closely to the primary signals.

VLF magnetic fields are detected by coils in which currents flow in proportion to core permeability, the number of turns and the field component along the coil axis. No signal will be detected if the magnetic field vector lies at right angles to this axis.

A VLF electric field induces an alternating current in an aerial consisting of a straight conducting rod or wire. The signal strength is roughly proportional to the amplitude of the electric-field component parallel to the aerial, and to the aerial length.

**Fig. 9.2** Major VLF transmitters. Data blocks list station identifiers (e.g. NAA), frequencies (kHz) and power (Megawatts). Frequencies and powers are liable to change without much notification; the frequency of station NAA was changed in 1985 and several surveys are known to have been subsequently attempted using the earlier frequency.

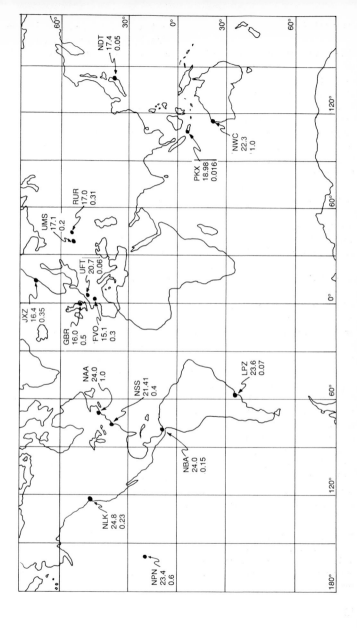

### 9.1.3 Magnetic field effects

Eddy currents induced by a VLF magnetic field in the more conductive parts of the ground produce secondary magnetic fields with the same frequency as the primary but generally with different phase. Any vertical magnetic component is by definition anomalous, and most VLF instruments compare vertical with horizontal magnetic fields either directly or by observing tilt angles.

Eddy currents in a near-vertical sheet-like conductor flow in rotational paths and produce secondary magnetic fields which oppose the primary field change at the body's surface. Directly above such a conductor the secondary field may be strong but will be horizontal and most systems will record no anomaly. On either side there will be detectable vertical fields, in opposite directions, producing an anti-symmetric anomaly (Fig. 9.3).

VLF anomalies are also produced by steeply dipping single contacts. If the further edge of a body is effectively beyond the area of investigation, the anomaly is either positive or negative, depending upon the sign convention (Fig. 9.4). The classical anti-symmetric 'thin conductor' anomaly can be looked upon as the limiting case of two contacts very close together.

Two steeply-dipping conductors close to each other produce a resultant anomaly generally similar to the sum of the anomalies that would have been produced by each body singly. Where, however, one of the bodies is steeply dipping and the other flat-lying, the results are more difficult to anticipate. Conductive overburden affects, and can actually reverse, the phase of the secondary field.

### 9.1.4 Electric field effects

Because the Earth is not a perfect conductor, VLF electric vectors are tilted, not vertical, near its surface, having horizontal components. Above homogeneous ground the horizontal field would differ in phase from the primary (vertical) field by 45°, would lie in the direction of propagation and would be proportional to the square root of the ground resistivity.

The phase angle will be greater than 45° if resistivity increases with depth and less than 45° if it decreases. The relationship is reversed if the horizontal magnetic field is used as the phase reference, since the electric and magnetic components of the primary field are 90° out of phase. Over a layered Earth, the tilt of the electric field records average (apparent) resistivity, strongly biased towards the resistivity of the ground within about half a skin depth of the surface. Sharp lateral resistivity changes distort this simple picture and very good (usually artificial) conductors produce secondary fields which invalidate the assumptions of the resistivity calculations.

### 9.1.5 Coupling

The magnetic-component response

126

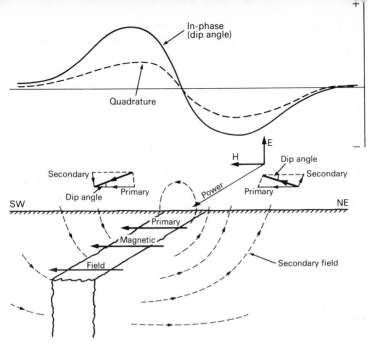

**Fig. 9.3** Magnetic-component anomaly over a vertical conducting sheet striking towards the transmitter. The magnetic field always dips away from the conductor and a sign convention is needed to decide which direction of dip is to be regarded as positive.

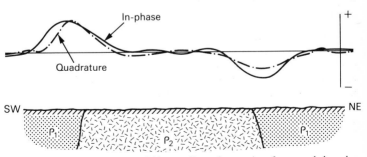

**Fig. 9.4** 'Contact' VLF magnetic-field anomalies at the margins of an extended conductor. Sign convention as for Fig. 9.3.

of a good conductor depends critically on its orientation. This is also true in conventional e.m. surveys but e.m. traverses are usually laid out at right angles to the probable geological strike, automatically ensuring good coupling. In VLF work the traverse direction is almost irrelevant, the critical parameter being the relationship between the strike of the conductor and the bearing of the transmitting station.

A body which strikes towards the transmitter is said to be *well-coupled*, since the magnetic vector is at right angles to it and current can flow freely. Otherwise current flow will be restricted, reducing the strength of the secondary field. If the probable strike of the conductors is either variable or unknown, two transmitters, bearing roughly at right angles, should be used to produce separate VLF maps.

A Mercator-projection map such as Fig. 9.2 is of only limited use in determining the true bearing of VLF transmitters; the all-important *Great Circle Paths* can be found using a computer program or a globe and a piece of string.

## 9.2 VLF instruments

The first commercially-available geophysical VLF instrument, the Ronka-Geonics EM-16, used only magnetic vectors, although horizontal electric fields can now be measured with the EM-16R add-on module. Other ground VLF instruments have come on to the market during the last few years, following expiry of the Ronka patents.

### 9.2.1 Elliptical polarization

Combining *horizontal* primary and secondary fields with different phase produces a *horizontal* field which differs from its components in both phase and magnitude. A secondary field which is *vertical* and *in phase* with the primary produces a resultant which has the same phase but is *tilted* and stronger. A vertical secondary field in *phase quadrature* with the primary results in *elliptical polarization* (Fig. 9.5).

These are special cases. In the general case of an inclined secondary field which is neither in phase nor in phase quadrature with the primary, a tilted, elliptically-polarized wave is produced. Because the secondary field has a horizontal component, the tangent of the tilt angle is not identical to the ratio of the vertical secondary field to the primary. Because of the tilt, the quadrature component of the vertical secondary field does not define the length of the minor axis of the ellipse. However, dip-angle data are usually interpreted qualitatively, when such distinctions, which are only significant for strong anomalies, are ignored. Quantitative interpretations are based on physical or computer model studies, the results of which can be expressed in terms of any quantities measured in the field.

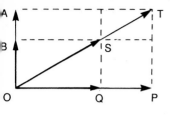

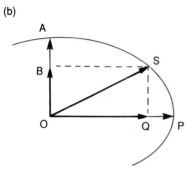

**Fig. 9.5** Combination of horizontal and vertical alternating fields. (*a*) In-phase: the vertical vector has its maximum value OA when the horizontal vector has its maximum value OP and the resultant then has its maximum value OT. All three are zero at the same time, and at any other time (represented by OQ, OB and OS) the resultant is directed along OT but has lower amplitude. (*b*) Phase-quadrature: the vertical vector is zero when the horizontal vector has its maximum value OP, and has its maximum value OA when the horizontal vector is zero. At other times, represented by OB, OQ and OS, the tip of the resultant lies on the ellipse as shown.

## 9.2.2 The EM-16

The EM-16 consists of a housing containing the electronic circuitry, to which is attached a conventional sighting clinometer, and a T-shaped handle containing two coils at right angles (Fig. 9.6). Controls include a two-position station-selector switch, a calibrated quadrature control and a knob which amplifies an audio tone which, despite being extremely irritating, can be very difficult to hear properly in areas such as forests on windy days, where other noises compete.

With the phase control at zero, the strength of the tone is determined by the setting of the volume control and by the component of the VLF magnetic field parallel to the axis of the main coil. Measurements are made by finding orientations of this coil which produce nulls (minima). This is easiest if the volume control is set so that at the null the tone is only just audible.

At each reading point the direction of the minimum horizontal component (i.e. the direction of the power vector) must first be determined. Unless there is a significant secondary field this also gives the bearing of the transmitter. The instrument is held with both coils horizontal, most conveniently with the short coil at right angles to the stomach (Fig. 9.7). The observer turns until he estimates that he has found a null, at which stage the magnetic field is at right angles to the main coil and parallel to the short coil. It is occasionally necessary to

**Fig. 9.6** The EM-16 in normal reading position.

**Fig. 9.7** Searching for the station with the EM-16 (Galicia, northern Spain).

adjust the quadrature control during this process; it should be re-set to zero before attempting to observe vertical field.

Without changing position, the observer then rotates the instrument *about the short coil as axis* into the upright position, bringing the clinometer up to eye level. Tilting in the plane of the clinometer will cause the signal to vary, with a minimum when the long coil is at right angles to the major axis of the polarization ellipse.

The null will be poorly defined if the quadrature component (the minor axis field) is large. Definition can be improved by using the quadrature control to phase-shift a measured percentage of the (major-axis) field detected by the short coil and subtract it from the quadrature field detected by the long coil. At the null, with the instrument held in the tilted position, the quadrature reading gives the ratio of the ellipse axes and the tangent of the tilt angle defines the in-phase anomaly.

The instrument is designed to be tilted with the clinometer dial in a vertical plane and a good null will be obtained only if the polarization ellipse is also vertical. If this is not the case, the null may be poorly defined, even after phase-shifting.

### 9.2.3 EM-16 sign conventions

Whether a tilt angle appears positive or negative depends on direction. If the conductor is to the front, the observer will have to lean backwards and will see a positive reading on the clinometer. The transmitter can be to either the left or the right, but to avoid confusion all readings in a survey should be taken facing in the same direction. This direction must always be recorded in the field notes even if, as is recommended practice, a standard range of directions (e.g. N and E rather than S or W) is specified for use on all surveys.

Reversed in-phase anomalies are caused either by steeply-dipping insulators enclosed in conductive country-rock, a rather rare circumstance, or by 'active' sources such as live power lines. Quadrature anomalies usually show the same sense of asymmetry as in-phase anomalies, but may be reversed by conductive overburden.

### 9.2.4 The EM-16R

With the additional circuitry contained in the EM-16R plug-in module and a two-metre length of shielded cable acting as an aerial, the EM-16 can be used to measure horizontal electric fields. The cable is stretched out on the ground towards the transmitter and the two ends are pegged down. For convenience, the instrument is usually also laid on the ground, with the long coil towards the transmitter; the short coil then detects the maximum magnetic-field component. A null is obtained by rotating the 16R control, giving a reading directly in ohm-metres. Phase shifts are also monitored.

Conversions of EM-16R readings, which use the horizontal magnetic field as a phase reference, to resistivities assume a fixed relationship between horizontal magnetic and vertical electric components. This condition is not satisfied where significant secondary magnetic fields are present and methods which reference to the more stable vertical electric field may be preferable.

### 9.2.5 Other VLF instruments

Most of the alternatives to the EM-16 also record magnetic vector variations but measure field components and their ratios rather than dip angles. Major advances include direct recording of data, often into solid-state memory as well as to a front-panel display, and elimination of the use of an audible tone. Some instruments can measure natural magnetic and two-transmitter VLF fields simultaneously, and some have been made self-orientating to increase speed of coverage.

Amplitude anomalies may also be measured, but primary fields are affected by meteorological changes along transmission paths and by variations in transmitter output; a second instrument is needed as a monitor. Horizontal magnetic-field directions are sometimes recorded but are generally less sensitive and less diagnostic than changes in tilt angle and require a directional reference, usually a small compass.

Most instruments rely, as does the EM-16, on crystal-controlled tuning to lock-on to the desired station. A few use high-Q tuning circuits, but with these there is often considerable

uncertainty as to which station is being acquired and nulls may be 'fuzzy' because of interference.

### 9.3 VLF maps

Dip-angle data can be awkward to contour and the resulting maps, on which conductors are indicated by steep gradients, may be difficult to assess visually. VLF results also tend to be rather noisy, minor anomalies being caused by small local (usually artificial) conductors. Large artificial conductors produce classic anti-symmetric anomalies but geological conductors are often indicated merely by sloping sections on profiles (Fig. 9.8).

### 9.3.1 Fraser filtering

The effects of noise can be reduced by adding together results recorded at closely spaced stations and treating the result as the value at the midpoint of the station group. This is the simplest possible form of low-pass filter. The asymmetry inherent in dip-angle data may be removed by differencing adjacent readings to obtain average horizontal gradients.

A filter designed to carry out both these operations was suggested by a Canadian geophysicist, Doug Fraser. Four equispaced consecutive readings are required to produce a single value. The first two readings are added together and the same is done with the latter two. The second average value is then subtracted from the first.

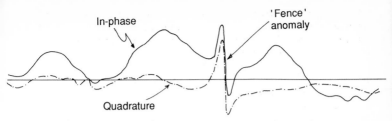

**Fig. 9.8** EM-16 profile in area of high geological noise levels, with superimposed anomaly due to a rabbit-proof fence.

The Fraser filter is widely used and some instruments can be programmed to apply it automatically. The results can usually be contoured quite easily, especially if, as is normal practice, negative values are discarded. Steeply-dipping conductors produce positive anomalies and are very obvious.

However, there are disadvantages. It is a truism of all geophysical work that processing degrades data. The Fraser filter can destroy subtle but possible significant features and,

### Example 9.1

A series of EM-16 readings begins:

$+2 +3 +6 +11 +11 +2 -7 -13$
$-8 -4 \dots$

The first Fraser filter value is:

$$(2+3)-(6+11)=-12$$

The second value is:

$$(3+6)-(11+11)=-13$$

The complete Fraser filter curve, together with the original VLF profile, is shown in Fig. 9.9

more importantly, distorts valid anomalies due to sources other than simple conductive sheets. For example, an isolated high or low indicative of a steeply-dipping interface between materials of different conductivity will be transformed by the filter into an anti-symmetric anomaly. If negative values are then ignored, this will inevitably be interpreted as due to a steeply-dipping conductor, offset from the actual zone of conductivity change.

### 9.3.2 Mapping VLF data

Since raw EM-16 data are difficult to contour and there are valid objections to Fraser filtering, stacked profiles offer the best presentation method. These display all the original data, correctly located on the map, and sections of profile on which there are abrupt gradients indicating a conductor can be emphasized by thickened lines.

In order for a map to be interpreted, even qualitatively, the direction to the transmitter must be shown so that the degree of coupling can be assessed. Conductors striking

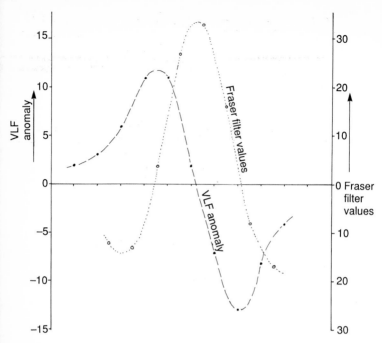

**Fig. 9.9** EM-16 profile and its Fraser-filter equivalent.

at right angles to this direction will not be well-coupled and may not be seen. The map must also show which of the two possible reading directions has been used, since it is otherwise not possible to distinguish between 'normal' conductors, and 'reversed' gradients which may be due to active sources such as power and telephone lines.

### 9.3.3 VLF/e.m. comparisons

VLF methods are best suited to mapping near-vertical contacts and fractures. Conductive mineralization may be detected, but the magnitudes of anomalies associated with very good conductors may be no greater than those produced by un-mineralized fractures. Many more anomalies will usually be located than by an e.m. survey over the same ground because VLF systems operate at the high-frequency end of the response curves, where most conductors appear strong (see Fig. 8.2); Fracture zones may not be as conductive as economic mineralization but are likely to occupy much more of the sub-surface. EM-16 and elec-

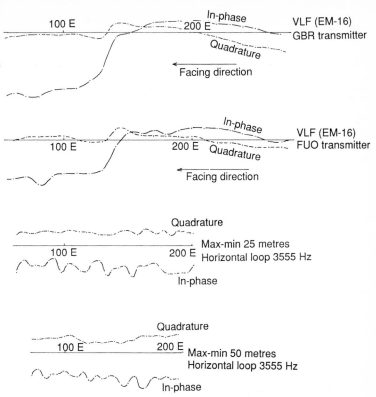

**Fig. 9.10** Comparison of EM-16 and conventional e.m. (horizontal-loop) results across a shear zone in granite. The in-phase variations on the Max–Min profiles are due to small spacing errors, which are more significant at the smaller coil separation. The source of the strong VLF anomaly was detected by the e.m. system only on the quadrature channel, indicating a poor conductor, and then only at the 50-m spacing using the highest frequency of which the instrument was capable. (Bavarian Forest; data courtesy BP Minerals International Ltd.)

tromagnetic profiles across a real conductor are compared in Fig. 9.10.

VLF measurements can be made quickly and conveniently by a single operator. They are therefore sometimes used to assess the elec-

tromagnetic characteristics of an area before the expense of a conventional e.m. survey is incurred. This is especially done near to populated areas where artificial sources are to be expected.

# 10

# Seismic methods—general considerations

Seismic methods are the most effect-ive, and by far the most expensive, of all geophysical techniques. Both reflected and refracted waves are used, and features common to both types of survey are considered in this chapter. Small-scale reflection work is further discussed in Chapter 11 and shallow refraction in Chapter 12. Deep reflection surveys are beyond the scope of this book since they involve large field crews, soph-isticated equipment and complex data processing.

## 10.1 Seismic waves

A seismic wave transmits energy by vibration of rock particles. Low-energy seismic waves can be regarded as elastic, leaving the rock mass unchanged after they have passed, but rock may be crushed and permanently distorted close to a seismic source.

### 10.1.1 Types of elastic wave

When a sound-wave travels in air, the molecules oscillate backwards and forwards in the direction of energy transport. Such a wave propa-gates as a series of compressions and rarefactions, being termed a pressure or 'push' wave. It has the highest velocity of any wave motion, and is therefore also known as the primary or simply the *P-wave*.

Rock particles can also vibrate at right-angles to the direction of energy flow, creating an *S-wave* (shear, 'shake' or, because of the rela-tively slow velocity, secondary). The velocity of an S-wave depends slightly on the plane in which the particles vibrate, but the differences are not important in small-scale surveys.

P and S wavefronts expand throughout the main mass of the rock and are therefore termed *body-waves*. At interfaces *Love waves* can be generated and at the Earth's surface *Rayleigh waves* can propagate, with particles following elliptical paths. Love and Rayleigh waves carry a considerable proportion of the source energy but travel very slowly and are of little geophysical use.

They are often simply lumped together as the 'ground roll'.

## 10.1.2 Seismic velocities

The 'seismic velocity' or simply 'velocity', of a rock is the velocity at which a wave motion propagates. It is quite distinct from the continually varying velocity with which individual rock particles oscillate.

Any elastic-wave velocity ($v$) can be expressed as the square-root of an elastic modulus divided by the square root of density. For P-waves the elongational elasticity ($j$) is appropriate, for S-waves the shear modulus ($\mu$):

$$v_P = \sqrt{\frac{j}{\rho}} \; ; v_S = \sqrt{\frac{\mu}{\rho}} \quad (10.1)$$

If density and these velocities are known, all the elastic constants of a rock mass can be determined since they are inter-related by the equations:

$$j = q\frac{1-\sigma}{(1+\sigma)(1-2\sigma)} \quad (10.2)$$

$$\mu = q/2(1+\sigma) \quad (10.3)$$

$$K = q/3(1-2\sigma) \quad (10.4)$$

$\sigma$ is the Poisson ratio, which is always less than 0.5, $q$ the Young's modulus and $K$ the bulk modulus. It follows that $j = K + \frac{4}{3}\mu$, and that a P-wave always travels faster than an S-wave in the same medium. In many consolidated rocks the P-wave travels roughly twice as fast.

Often only P-wave velocities can

be measured, providing a rather rough guide to rock quality. Figure 10.1 shows ranges of velocity for common rocks and also of 'rippability', the extent to which a rock can be ripped apart by a spike mounted on the back of a bulldozer.

## 10.1.3 Velocities and the time-average equation

Clays and other weathering products have low seismic velocities and high porosities, and weathered rocks therefore have lower velocities than the same rocks when fresh. This fact lies behind the rippability ranges of Figure 10.1; few fresh, consolidated rocks have velocities less than about 2200 m/sec, and rippable rocks are generally at least partly weathered.

Within quite broad limits, the velocity of a mixture can be obtained by averaging, not the velocities of the two constituents, but transit times weighted according to the relative proportions of the two present. The principle can even be used when one of the constituents is a fluid (Example 10.1).

The same sandstone, if dry, would have air with velocity of 330 m/sec in the pore spaces and the P-wave velocity would be much lower, although time-averaging cannot be used quantitatively for gas-filled pores. Many dry materials have very low P-wave velocities and absorb S-waves because they are poorly consolidated and do not respond elastically. Wet, poorly consolidated

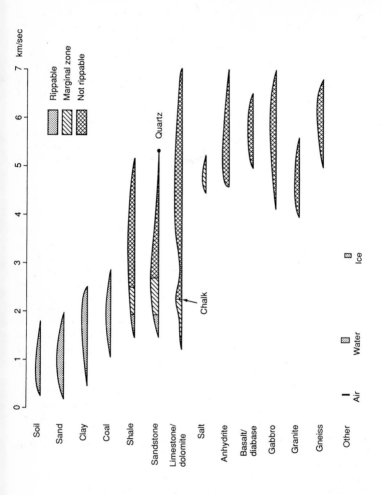

**Fig. 10.1** Ranges of P-wave velocity for common rocks. 'Rippability' is indicated by shading.

materials have velocities a little greater than that of water, and the top of the water table is often a prominent seismic interface.

## Example 10.1

---

P-wave velocity in quartz:
$$5200 \text{ m/sec}$$
P-wave velocity in water:
$$1500 \text{ m/sec}$$

The velocity in a sandstone, 80% quartz, 20% water-filled porosity, is given by:

$$\frac{1}{v} = \frac{0.8}{5200} + \frac{0.2}{1500} = 0.000287$$

$$v = 3480 \text{ m·sec}^{-1}$$

---

### 10.1.4 Ray-path diagrams

Only a small part of a wavefront is of interest in a geophysical survey since only a small part of the energy returns to the surface at points where detectors have been placed. It is convenient to identify the important travel paths by drawing seismic 'rays' to which the laws of geometrical optics can be applied.

Useful seismic waves have wavelengths generally between 25 and 200 m and ray-path theory works less well in seismology than in optics because of these large values. However, field interpretations can still be based on ray approximations.

### 10.1.5 Reflection and refraction

When a seismic wave encounters an interface between two different rock types, some of the energy will be reflected and the remainder will continue on its way at a different angle, i.e. it will be refracted. The law of reflection is very simple; the angle of reflection is equal to the angle of incidence (Fig. 10.2a).

Refraction is governed by *Snell's Law*, which relates the angles of incidence and refraction to the seismic velocities in the two media:

$$\frac{\sin i}{\sin r} = \frac{v_1}{v_2} \tag{10.5}$$

Refraction will be towards the interface if $v_2$ is greater then $v_1$ (Fig. 10.2b). If $\sin i$ equals $v_1/v_2$ the refracted ray will be parallel to the interface and some of the energy will return to the surface after leaving the interface at the original angle of incidence. This is the basis of the refraction methods discussed in Chapter 12. At greater angles there can be no refracted ray and all the energy is reflected.

When drawing accurate ray paths, allowance must be made for refraction through all shallower layers. Only the normal-incidence ray, which meets all interfaces at right angles, is not affected.

## 10.2 Seismic sources

The traditional seismic source is a small charge of dynamite. Impact and vibratory sources are now more

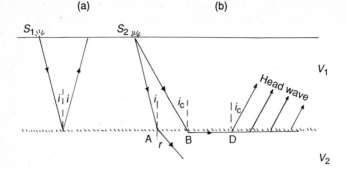

**Fig. 10.2** (*a*) Reflection. (*b*) Refraction. At A, where simple refraction is taking place, $\sin i/\sin r = v_1/v_2$. At B, where there is critical refraction, $\sin i_c = v_1/v_2$. Energy leaves the interface from points such as D to form the planar *head-wave*, which returns to the surface.

popular but explosives are still in quite common use.

### 10.2.1 *Explosives*

Almost any type of (safe) explosive can be used for seismic work, particularly if the charges are to be buried in shallow holes and will not be subject to unusual temperatures or pressures. Cord explosive, used in quarry blasting to introduce delays into a firing sequence, is rather safer to handle than normal gelignite; shot-holes can then be prepared by driving a metal rod or crowbar into the ground, feeding in the cord and tamping down. Single detonators can be used for shallow reflections.

Although explosives deliver a great deal of energy, much may be wasted in shattering rock in the immediate vicinity of the shotpoint. More of the energy is converted into seismic waves if the shot is fired in a metre or so of water. This effect is so marked that, if shot position is not critical, it can be worth going tens or even hundreds of metres further from the recording spread in order to put the charge in a river. Significant improvements can be obtained by watering shot holes in dry areas.

Electrical firing is normal when using explosives but with ordinary detonators there is a short delay between the instant at which the filament burns through, which provides a time reference, and the time at which the main charge explodes. Special 'zero-delay' detonators should be used for seismic work. Delays should be routinely checked by firing a single detonator buried a few inches from a geophone and recording the result. This measures total delay through the whole system including the recorders.

Explosives involve problems not only with safety but also with bureaucrats and security. Gelignite and detonators must normally be stored in separate secure and licensed premises and must be used in conformity with local regulations. In some countries the work must be supervised by a licensed shot-firer, and police permission is required almost everywhere. Despite these disadvantages, and despite the headaches which are instantly produced if gelignite comes into contact with bare skin, explosives are still used; they represent potential seismic energy in its most portable form and are virtually essential in rugged terrain if signals must be detected at distances of more than 20–30 m.

Seismic waves can be generated by devices which fire lead slugs into the ground from shotgun-type cartridges. The energy supplied is relatively small, and a firearms certificate may be needed, at least in the UK. Another approach is to use blank shotgun cartridges in a small auger which incorporates a firing chamber, combining the shot-hole and the shot. Even this seldom provides more energy than a blow from a well swung hammer, and repetition is less simple.

## 10.2.2 Hammers

A 4- or 6-pound sledgehammer provides a versatile source for small-scale surveys. The useful energy produced depends on ground conditions as well as on strength and skill. Hammers can nearly always be used in refraction work on spreads 10 to 20 m long but very seldom where energy has to travel 50 m or more.

The hammer is aimed at a flat 'plate', the purpose of which is not so much to improve the pulse (hitting the ground directly can sometimes provide more seismic energy) but to stop the hammer abruptly and so provide a definite and repeatable shot instant. Inch-thick aluminium or steel plates used to be favoured, but are now being replaced by rubber discs which last longer and are much less noisy.

The first few hammer blows are often rather ineffective, as the plate needs to 'bed down' in the soil. Too much enthusiasm may later embed it so deeply that it has to be dug out.

## 10.2.3 Other impact sources

More powerful impact sources can be used in larger surveys. Weights of hundreds of kilograms can be raised by portable hoists or cranes and then dropped (Fig. 10.3). The minimum release height is about 4 m, even if a shorter drop would provide ample energy, since rebound of the support when the weight is released creates its own seismic wavetrain. A long drop allows these vibrations to die away before the true impact occurs.

Tractor-mounted post-hole drivers, in common use in many agricultural areas, make convenient

**Fig. 10.3** Impact source. A half-ton weight being dropped from a portable crane in the course of a low-velocity-layer survey near Mildura, Australia.

sources. The weight drops down a guide and is raised by a pulley system connected to the tractor power take-off.

Use is sometimes made of a relatively small (70 kg) weight falling in an evacuated tube. The upper surface is exposed to the atmosphere and effectively several hundred extra kilograms of atmospheric column are also dropped. The idea is elegant but the source is difficult to transport in rugged terrain because the tube must be strong and therefore heavy and must be mounted on a trailer. Moreover, a motor compressor is needed to pump out the air, adding to the weight and complexity of the system.

### 10.2.4 Time-breaks

In any seismic system the time at which the seismic wave is initiated must be known. In some instruments this appears on the record as a break in one of the traces (the shot- or time-break); on others it actually defines the start of the record.

Time-breaks may be generated in many different ways. A geophone may be placed close to the source, although this is very hard on the geophone. Explosive sources are usually fired electrically, and the cessation of current flow in the detonator circuit can provide the required signal. Alternatively, the time-break circuit can include a loop wound round the main explosive charge, to be broken at the shot instant. This technique may be used in those rare cases where charges are fired using lit fuses.

The time-break on a hammer usually involves making rather than breaking a circuit. A relay on the back of the handle, just behind the head, will close momentarily when the hammer stops on striking the plate (Fig. 10.4). It will close late or not at all if the hammer is used the wrong way round. Solid-state switches give more repeatable results but are expensive and rather easily damaged.

Contact can also be made if the hammer head is connected to one side of the circuit and the (metal) plate to the other. Although this sounds a simple and foolproof technique, in practice the repeated shocks suffered by the various connections are too severe for long-term reliability. The plates themselves have rather short lives, after which new connections have to be made.

However the time-break is achieved when using a hammer, there will always be a cable linking it to the recorder. This may snake across the plate just before impact and be cut. If this should happen, the culprit has both to repair the damage and buy drinks for all the witnesses!

Where higher-power impacts are obtained by dropping heavy weights from considerable heights, relay switches can be attached to the top surfaces of the weights but are unreliable unless an absolutely straight drop can be guaranteed. A rather crude but effective home-made device which can be attached to any dropping weight is shown in Fig. 10.5.

Time-break pulses are often rather strong and can produce considerable interference on other channels ('cross-talk'; see Section 10.3.5). Time-break cables and circuits are therefore usually separated from data lines.

**Fig. 10.4** 'Post-office relay' impact switch on the back of a sledgehammer handle.

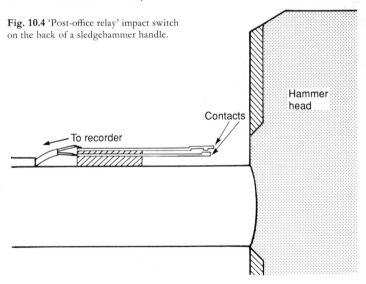

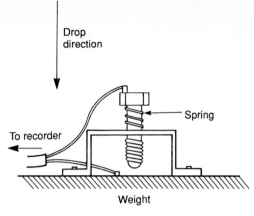

**Fig. 10.5** Weight-drop contact switch. On impact, the inertia of the bolt compresses the spring, making contact with the upper surface of the weight.

*10.2.5 Safety*

Large amounts of energy must be supplied to the ground if refractions are to be observed from depths of more than a very few metres, and such operations are inherently dangerous. Risks are greatest when explosives are being used, but it is no safer to stand beneath a half-ton of steel being dropped from a height of 4 m.

The basic safety principle is that the shot-firer must be able to see the shot point. Unfortunately, some seismographs are so designed that the shot is fired by the instrument operator, who can seldom see anything and who will in any case be preoccupied with monitoring noise levels. It must at least be possible for firing to be prevented by someone who is far enough from the shot to be safe but close enough to see what

is happening. With explosives the detonator can be linked to the firing unit indirectly via 20 or 30 m of secondary cable. This should be connected to the detonator after the shot hole has been charged, and to the cable from the power source only when the shot point is clear. Disaster can be prevented at any time by pulling these two cables apart.

Unless 'sweaty' gelignite is being used (and the sight of oily nitroglycerine oozing out of the packets should be sufficient warning to even the least experienced), modern explosives are reasonably insensitive to both heat and shock. Small amounts of gelignite can, for example, be burnt on an open fire and this is a recommended method for disposing of inconvenient surpluses. Detonators are the usual causes of any serious safety problems.

It is possible, even if unlikely, for a detonator to be triggered by currents induced by power-lines or radio transmissions. The likelihood of this happening is reduced if the detonator leads are twisted together. The possibility of initiation by static electricity is eliminated if the circuit is closed. The free ends of the leads should be parted only when the time comes to connect to the firing cable, which should itself be shorted out at the far end. Explosives should not be handled when thunderstorms are about.

Although the explosive power of a detonator on its own is rather small, it must still be treated with respect. Fingers and even hands have been lost in detonator explosions. If fired on their own, detonators should always be placed in well-tamped holes, since damage or serious injury can be caused by fragments of the metal casing.

Explosive charges need to be adjusted to the holes available. Large charges may be used in deep holes with little obvious effect at the surface, but a hole less than 2 m deep will often 'blow-out', scattering debris over a wide area. Only experience will allow safe distances to be estimated, and even experienced users can make mistakes; safety helmets should be worn and physical shelter such as a wall, a truck or a large tree should be available. Heavy blasting mats can reduce blow-outs, but their useful lives tend to be short and it is unwise to rely on them alone.

A point where a shot has been fired but no crater has formed must be regarded with suspicion. The subsurface cavity may collapse at some later date under the weight of a person, animal or vehicle, leading to interesting litigation.

## 10.3 Detection of seismic waves

Land seismic detectors are known as geophones, marine detectors as hydrophones.

### 10.3.1 Geophones

A modern geophone consists of a coil wound on a high-permeability magnetic core, spring-suspended in the field of a permanent magnet (Fig. 10.6). If the coils move relative to the magnet, voltages are induced and current flows in any external circuit. The current is proportional to the velocity of the coil through the magnetic field, so that ground movement is recorded, not ground displacement.

Geophones are usually positioned by pushing a spike screwed to the casing firmly into the ground; it may be necessary to unscrew the spike and use some form of adhesive pad or putty in surveys carried out over bare rock.

The relative motion of coil and casing is influenced by the natural vibration frequency of the suspended system and by the degree of damping. At frequencies above resonance, the response will approximate to that of an ideal system,

**Fig. 10.6** Moving-coil geophone 'cut-away' to show coil. Centimetre scale-bar on plinth indicates size. (Courtesy, Sensor Nederland B.V.).

surface interfaces. For large offsets in reflection work, or where refractions are occurring at interfaces where velocity contrasts are small, the wavefronts will be inclined to the ground surface and the discrimination between S- and P-waves will be less good.

Geophones are remarkably rugged, which is fortunate considering the ways in which they are often treated. Their useful life is certainly reduced if, as is common practice, they are dumped unceremoniously from trucks into tangled heaps on the ground. Frames can be bought or made to which they can be clipped for carrying, and these are certainly a good investment if actually used (Fig. 10.8).

### 10.3.2 Detection of S-waves

Although S-waves are merely noise in most seismic work, there are occasions when S-wave information is specifically sought. In particular both S-wave and P-wave velocities are needed to determine elastic properties (see Section 10.1.2).

The coil in an S-wave geophone is free to move in the horizontal rather than the vertical plane. It is assumed that the wavefront rises more or less vertically, so that the S-wave vibrations are in the plane of the ground surface. Where direct waves travelling parallel to the surface are concerned, 'S-wave' geophones are in fact more sensitive to P-waves and 'P-wave' geophones to vertically-polarized S-waves.

faithfully replicating the ground motion. Near and below resonance, the signal will be heavily attenuated. Industry-standard geophones resonate at about 10 Hz, which is well below the frequencies useful in small-scale studies. Response curves for a typical 10 Hz phone are shown in Fig. 10.7. Normal geophone coils have resistances of the order of 400 ohms and damping is largely determined by the impedance of the circuit to which the geophone is linked.

Geophones are generally designed to be mounted with the coil free to vibrate vertically, since they are then most sensitive to reflected and refracted P-waves rising steeply from sub-

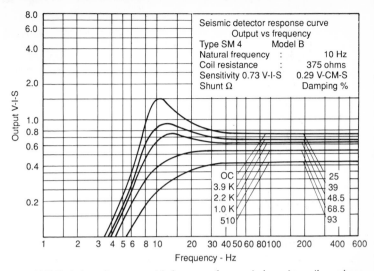

**Fig. 10.7** Variation of response with frequency for a typical moving-coil geophone. Degree of damping depends largely on the input resistance of the instrument. (Reproduced by courtesy of Sensor Nederland B.V.).

**Fig. 10.8** Geophone carrying-frame, in use, Papua New Guinea.

### 10.3.3 Detection in swamps and in water

Normal geophones are rainproof rather than waterproof, and are connected to cables by open crocodile clips. For use in swamps, geophones are available which are completely enclosed and sealed into waterproof cases. These do not have external spikes but are shaped so that they can easily be pushed into mud.

Movement-sensitive instruments are ineffective in water, where geophones must be replaced by *hydrophones* which respond to variations in pressure rather than motion and are equally sensitive in all directions. Discrimination between P- and S-waves is not required since S-waves cannot propagate in liquids.

### 10.3.4 Noise

Any vibration which is not part of the signal is termed *noise*. Noise is inevitable, and indeed some noise, termed coherent, is generated by the shot itself. S-waves, surface waves and reflections from near-surface irregularities are all forms of coherent noise and little can be done in the field to reduce their effect. In shallow refraction work the problems presented by these slow waves are usually avoided by using only the times at which energy first arrives at each geophone.

Noise which is not generated by the shot is termed *random*. Movements of traffic, livestock and people are all sources of random noise and, to varying extents, can be controlled.

It should at least be possible to prevent members of the survey team contributing. Some form of whistle or hooter, warning of the imminence of a shot or impact, can be invaluable.

Random noise is also produced by wind moving vegetation and so disturbing the ground. Some improvement can be achieved by siting geophones away from trees and bushes, and sometimes by clearing smaller plants. The actual placement of geophones can also be important; in hard ground the mounting spike is hard to push fully home, but a geophone an inch above the ground vibrates very readily. Often one or two geophones are particularly noisy, and major improvements can sometimes be made by moving the offending units a few inches.

### 10.3.5 Seismic cables

Seismic signals travel from geophones to recorders as varying electric currents, and the cables carrying them must contain twice as many individual wires as there are geophones. The wires are necessarily packed very close together and not only can external current-carriers such as power and telephone cables induce currents, but a very strong signal in one channel can be passed inductively to all the others. This phenomenon, known as *cross-talk*, is particularly severe in the case of high amplitude signals from geophones close to the shot point, and it may even be necessary to disconnect these so as to obtain good records from distant detectors.

The amount of cross-talk varies but generally increases with cable age, probably because of a gradual build-up of moisture within the outer insulating cover. Eventually the cable must be discarded.

Cables and plugs are the most vulnerable parts of a seismic system and cables are particularly at risk where they enter plugs. Great care should be taken at all times; re-soldering wires to a plug with 24 or more connections is neither easy nor interesting.

Most cables are double-ended and either end can be connected to the receiver. It is quite common for one or more of the wires to be broken and for the corresponding channel to be 'dead'. If this happens, it will usually only affect the connection to one of the two ends, and it is worth reversing the cable to see if better results can be obtained. All too often the only result of doing this is the discovery of other dead channels.

## 10.4 Recording seismic waves

Instruments which record seismic waves are known as *seismographs*. They range in sophistication from timers which measure only single time intervals to complex microprocessor-controlled units which digitize, filter and record signals from a number of detectors simultaneously. Most modern instruments have temporary displays from which data can be written down in the field; most also provide for the production of 'hard-copy', either by dot-matrix printer or photographically.

### 10.4.1 Seismic timers

Simple timing circuits can be switched on by the time-break and off by the signal from a geophone. Single-channel instruments of this type are light and very portable; their limitations can be understood by considering a record produced by a true

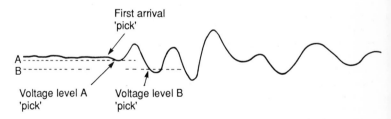

**Fig. 10.9** Typical single-channel seismic trace, hammer source. An automatic pick at a pre-set voltage level can clearly not occur as early as the visual estimate, made at a point where the voltage level is close to zero. If the trigger level is below A, or the trace is amplified less, a later event will be picked. If the trigger level is too close to zero, a noise pulse may forestall the true first arrival.

single-channel seismograph (Fig. 10.9). The trace consists of a relatively flat portion where only noise is being recorded, followed by a portion on which there is a strong signal. An arrival time has been 'picked' by eye at the point where the turn-down to the first signal trough appears to begin. This would become progressively more difficult to identify as source–detector distance increased. Amplification helps to some extent but does not improve the signal-to-noise ratio.

Seismic timers rely on timing circuits being switched off by input voltages which, after amplification, exceed a certain level. Such triggering can obviously not take place as early as the time picked on Fig. 10.9, as the voltage level is then close to zero. The measured time will vary with trigger setting and with amplification.

Timings can be totally incorrect. If the trigger level is too low, the count may be stopped by random noise before any signal arrives; too high a level and the first trough may be missed altogether. Triggering may then occur at some later peak, since later cycles tend to carry more energy.

'Shots' should be repeated a number of times, preferably using different trigger settings or amplifications, to confirm results. Noise levels can be estimated by starting the count without using a source (e.g. by closing the contacts on the relay switch) and seeing how long it is before counting is stopped by a noise pulse.

Seismic timers are very crude instruments and only worth using at all with sledgehammers, the cheapest of all sources. Only first-arrivals can be detected and so the only possible applications are in shallow refraction work. Even there, cheapness and portability compensate inadequately for the uncertainty which always surrounds the interpretations.

### 10.4.2 Single-channel seismographs

Most single-channel seismographs now use cathode-ray tube displays. The time range is set by front-panel switches and the left-hand edge of the screen marks the shot or impact instant. Hard-copy is not usually obtainable and times are measured directly. In some models a cursor can be moved across the screen, the time corresponding to its position being displayed on the front panel. Noise levels can generally be observed by use of a switch which allows continuous display of the signal without triggering the source.

Some single-channel instruments now use enhancement principles. A digital version of the signal is stored in solid-state memory, as well as being displayed on the screen. A second signal can either replace this or can be added to it. Any number $n$ of signals can be summed in this way, for a theoretical improvement in signal-to-noise ratio equal to $\sqrt{n}$.

Single-channel seismographs which allow signals to be displayed and summed are obviously superior to simple timers, and can be used to study events other than first arrivals. However, they are generally used only in shallow refraction work since it is difficult to distinguish between direct waves, refractions and reflections on a single trace. Hammer sources are universal, since using an explosion to obtain one trace only would be costly and inefficient.

### 10.4.3 Multi-channel seismographs

Seismographs with 12 or 24 channels are now common in shallow surveys (48 or 96 channels are the norm in deep reflection work). Enhancement is used very widely and most instruments produce both cathode-ray tube displays and optional hard-copy. The refraction survey example in Fig. 10.10 shows the signals recorded by 12 geophones at points successively further from the source. The traces from further geophones have been amplified more to compensate for attenuation. Multiple channels allow both refraction and reflection work to be done and explosives can reasonably be used since the cost per shot is less important when each shot produces many traces.

An enhancement seismograph (Fig. 10.11) is a very sophisticated and versatile instrument. Display formats can be varied, individual traces can be selected for enhancement, replacement or preservation; amplification can take place, ideally after as well as before storage in memory, and filters can be used. Time offsets can be used to display events occurring after long delay times. An important use of filters can be to reduce long-period noise of uncertain origin which sometimes drives the traces from one or two geophones across the display and obscures other traces.

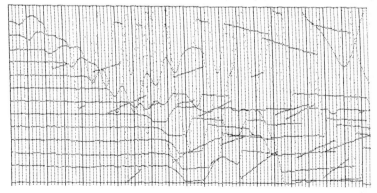

**Fig. 10.10** A 12-channel dot matrix record (EGG-Geometrics Nimbus 1210F), showing shallow refractions. Timing lines are at 1-msec intervals.

**Fig. 10.11** A modern enhancement seismograph (ABEM Terraloc).

Arrival times can be estimated from the screen while in the field but this is never easy and seldom convenient. On the other hand, hard-copies, especially if produced by dot-matrix printers, are often of rather poor quality, the matrix size causing irregularities in what should be smooth curves. The ideal, but uncommon, approach is to both estimate times in the field and retain hard-copy for later analysis. Grid lines can be superimposed elec-tronically on screens to aid estimation, but since the lines are not numbered, a hand-made scale on a piece of card is also needed.

Since enhancement requires digital storage of signals, it is rela-tively simple to arrange for digital output. Most instruments now incor-porate this option; the most con-venient are those which can output data to an ordinary portable cassette recorder.

# 11

# Seismic reflection

More than 90% of the money spent world-wide on applied geophysics goes into seismic reflection surveys, most of which aim at defining structures down to depths of thousands of metres using hundreds or even thousands of detectors. However, some reflection work is done by small field crews using relatively simple instruments to probe to depths of, at most, a few hundred metres.

## 11.1 Reflection theory

Ray path diagrams provide useful insights into the timing of reflection events. They give no indication as to amplitude.

### 11.1.1 Reflection coefficients and acoustic impedances

The *acoustic impedance* of a rock is equal to its density multiplied by its seismic velocity. If a seismic wavefront strikes a planar interface between rock strata with impedances

$I_1$ and $I_2$ at normal incidence, the percentage of energy reflected (the *reflection coefficient*, RC) is given by:

$$RC = \frac{I_2 - I_1}{I_2 + I_1} \qquad (11.1)$$

If $I_1$ is greater than $I_2$, the coefficient is negative and the wave will be reflected with phase reversed, i.e. a negative pulse will be returned where a positive was transmitted and vice-versa.

The proportion of energy reflected increases as the angle of incidence increases, until ultimately, if the velocity is greater in the second medium than in the first, there is total reflection and no transmitted wave. However, most small-scale surveys use waves reflected at nearly normal incidence.

### 11.1.2 Normal moveout

The true normal-incidence ray cannot be used in survey work, since a detector at a shot point would probably be damaged and would certainly be set into such violent oscillation that the whole record would be unusable. Geophones are therefore offset from sources and travel times must be corrected geometrically.

Figure 11.1 shows reflection from a horizontal interface, depth $a$, to a geophone offset a distance $x$ from the source. The travel time $T$ is related to the normal incidence time $T_0$ by the hyperbolic equation:

$$T^2 - T_0^2 = x^2/v^2 \qquad (11.2)$$

For small offsets, equation (11.2) can be replaced by the approximate parabolic form:

$$T - T_0 = x^2/(2v^2 T_0) \qquad (11.3)$$

The quantity $T - T_0$ is known as the *normal moveout* (NMO). Since velocity usually increases with depth and travel time always does, NMO decreases with depth.

Curved line-ups of reflection events can be seen on many multi-channel records (Fig. 11.2). If the source is at the centre of the geophone spread, the curves obtained over horizontal interfaces are symmetrical about this point. Curvature is the most reliable way of distinguishing shallow reflections from refractions. For deeper horizons the critical distance for refraction will usually lie beyond the furthest geophone.

### 11.1.3 Dix velocity

If there are several different layers above a reflector, the NMO equation will give the 'root-mean-square' (r.m.s.) velocity defined as:

$$v^2_{\text{r.m.s.}} = \frac{(T_1 v_1^2 + T_2 v_2^2 + \ldots T_n v_n^2)}{(T_1 + T_2 + \ldots + T_n)} \qquad (11.4)$$

where $T_n$ is the transit time through the $n$th layer, velocity $v_n$.

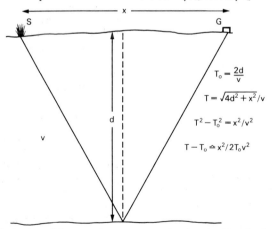

$$T_0 = \frac{2d}{v}$$

$$T = \sqrt{4d^2 + x^2}/v$$

$$T^2 - T_0^2 = x^2/v^2$$

$$T - T_0 \approx x^2/2T_0 v^2$$

**Fig. 11.1** Derivation of the normal moveout equation for a horizontal reflector. The normal-incidence time $T_0$ is equal to $2d/v$.

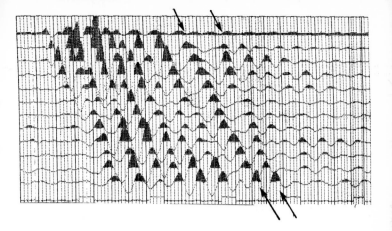

**Fig. 11.2** Enhancement seismograph record showing curved line-ups of reflectors (arrowed). Earlier events are produced by direct waves and refractions. The 'variable area' presentation used here is popular in reflection work since it emphasizes trace to trace correlations, although with some loss of information where overlap occurs.

Interval velocities can be calculated from r.m.s. velocities by using the *Dix Formula*:

$$v^2_{\text{int}} = \frac{(v_b^2 T_b - v_a^2 T_a)}{(T_b - T_a)} \qquad (11.5)$$

The subscripts a and b denote the times and r.m.s. velocities to, respectively, the top and bottom of the interval. Using r.m.s. velocities as if they were conventional average velocities can lead to significant errors in depth estimates.

### 11.1.4 Dip moveout

If a reflector has a uniform dip α, the reduction in travel path on the up-dip side of the shot compensates, up to a point, for the offset and travel times shorter than the normal-incidence time are possible (Fig. 11.3). The minimum time $2d\cos\alpha/v$ is recorded at the point, offset $2d\sin\alpha$ on the updip side of the shot, at which the reflected ray rises vertically to the surface. The moveout curve is symmetrical about this point. In shallow surveys these differences are detectable only for very large dips or very long spreads.

### 11.1.5 Multiple reflections

A wave reflected upwards with high amplitude from a sub-surface interface can be reflected downwards again from the ground surface and then back from the same interface. This is a simple multiple. Two strong reflectors can generate intraformational and peg-leg multiples as shown in Fig. 11.4.

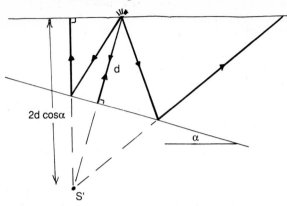

**Fig. 11.3** Rays are reflected from a dipping interface as if derived from the image point S' at depth $2d.\cos\alpha$ below the surface (d is the perpendicular distance from the source to the interface). The normal incidence travel time is $2d/v$, but the shortest travel time, $2d.\cos\alpha/v$, is associated with the ray that rises vertically back to the surface after reflection. An identical moveout hyperbola would be produced if the interface were horizontal and at a depth of $d.\cos\alpha$ and the shot point were offset a distance $2d.\sin\alpha$ in the updip direction.

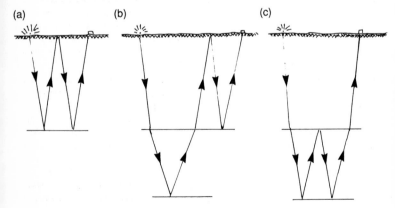

**Fig. 11.4** Multiple reflections. (*a*) Simple multiple. (*b*) Peg-leg (one leg shorter than the other). (*c*) Intra-formational multiple.

156

Multiples can usually be recognized on multi-channel records because they have moveouts appropriate to shallow reflectors and simple time relationships with their primaries. They are difficult to identify with certainty on single traces.

## 11.2 Reflection surveys

Reflected waves are never first arrivals, so clear-cut reflection events are seldom seen. Oil-industry techniques for improving signal-to-noise ratios are also available for shallow work, but are often too expensive to be used.

### 11.2.1 Spread lengths

Spreads for reflection work are usually much shorter than those used in refraction surveys to probe to similar depths. The distance from the source to the nearest geophone is usually dictated by the strength of the source, and may be as little as 2 m when a hammer is being used. Even with explosives or heavy weight drops, minimum offsets of more than about 10 m are unusual in shallow work.

With powerful sources and multi-channel recording, the furthest geophone may be more than 100 m from the source. The best spread length can be determined only by experiment, since the most important factors are the arrival times of the direct wave and any strong refracted waves. Field work should begin with tests specifically designed to examine these arrivals, generally by using elongated spreads.

### 11.2.2 Arrays

Ideally, reflected energy should arrive after the near-surface waves (ground-roll and refractions) have passed but this may not be possible if the depth of investigation is very small. In such cases, arrays of geophones may be connected to each recording channel. Reflected waves, which travel almost vertically, will reach all the geophones in an array at almost the same time but the direct waves will arrive at different times and produce signals which can interfere destructively.

The efficiency with which a wave is attenuated is defined by the *relative effect* (RE) of the array, compared to the effect of the same number of geophones placed together at the array centre. The variation with frequency of the RE of a linear array of five geophones, equally-spaced along a line directed towards the shot point and centred on the theoretical 'single-geophone' position, is shown in Fig. 11.5.

Simple arrays are preferred in the field, since mistakes are all too often made in setting out complicated ones. Using more than five geophones per array would be unusual in a shallow survey. Long arrays attenuate direct waves over a greater frequency range and it is often necessary to overlap the geophones in adjacent arrays.

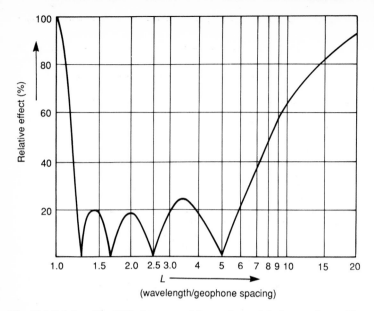

**Fig. 11.5** Relative effect (RE) of an array of five equi-spaced, in-line geophones. The 100% effect is equivalent to the effect of placing all the geophones at one point. Attenuation is concentrated between $L$ values of about 1.2 and 7. With 2-m spacing, a 500-m/sec direct wave would be attenuated at frequencies of between about 35 and 200 Hz.

### 11.2.3 Shot arrays

Seismic cables with only 12 or 24 channels are not designed with arrays in mind, and non-standard connectors may have to be fabricated to link the geophones to each other and to the cable. It may be easier to use shot arrays instead.

A shot array using explosives usually involves simultaneous detonation of charges laid out in a pattern resembling that of a conventional geophone array. If an impact source is used with an enhancement instrument, the same effect can be obtained by adding together results obtained with the impact at different points. This is the simplest way of attenuating surface waves from a hammer.

### 11.2.4 Common mid-point shooting

Stacking, the technique of improving signal-to-noise ratios by adding together several traces, is fundamental to deep reflection surveys. Enhancement seismographs in

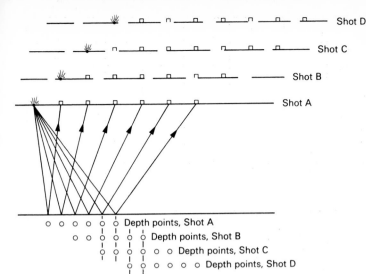

**Fig. 11.6** Schematic of field operations for obtaining 3-fold cover with a 6-channel seismograph. Successive shot points A, B, C, and D are placed progressively one geophone interval further to the right. The distance between reflection points on the subsurface interface (the depth points) is only half that between geophones on the surface. Shots A and D thus have no depth points in common.

shallow surveys are normally used to stack results obtained with identical source and detector positions, but if the data can be recorded on magnetic tape, corrections can later be made to traces produced with different source–receiver combinations, allowing them to be stacked.

The most popular system of stacking is to collect together a number of traces which have the same mid-point between source and receiver. NMO corrections must be applied to these *common midpoint* (CMP) traces before stacking, and processing is therefore not done in the field. If the interface dips, the velocity deduced from a CMP stack is equal to $v/\cos \alpha$

and the depth is the normal incidence distance from the mid-point. In contrast to the field set-up considered in Section 11.1.4, in which only a single shot point is used, the minimum time for a CMP gather is associated with the normal-incidence ray.

The number of traces gathered together in a CMP stack defines the 'fold' of coverage. Three traces gathered together to form a single synthetic zero-offset trace constitute a 3-fold stack, providing 300% cover.

The maximum fold obtainable, unless the shot point and geophone line are moved together by fractions of a geophone interval, is equal to half the number of data channels.

Figure 11.6 shows the successive geophone and source positions when a six-channel instrument is used to obtain 300% cover. Special cables and switching circuits are used in deep reflection CMP work, but with the instruments available for shallow surveys the method is clumsy and not often used.

CMP stacking was formerly known as *common depth point* (CDP) stacking, the assumption being that all the reflections would come from the same point on the sub-surface interface. This is only strictly true for horizontal interfaces and the newer term is preferable.

## 11.2.5 Depth conversion

Reflection events are recorded not in depth but in *two-way time* (TWT). Velocities are needed to convert times into depths, but those obtained from NMO curves may be 10–20% in error. Interpretations should be calibrated against borehole data wherever possible, and field crews should always be on the lookout for opportunities for measuring vertical velocities directly.

# *Seismic refraction*

Refraction surveys are widely used to study the water table and, for engineering purposes, the poorly consolidated layers near the ground surface, and also in determining near-surface corrections for deep reflection work. Travel times are usually of the order of only a few tens of milliseconds and there is little separation between arrivals of different forms of wave or of waves which have travelled by different paths. Only the first arrivals, which are always of a P-wave, can be 'picked' with any degree of confidence.

## 12.1 Refraction surveys

Ideally the interfaces studied in a small refraction survey should be shallow, roughly planar and dip at less than 15°. Velocity should increase with depth at each interface; the first arrivals at the surface will then come from successively deeper interfaces as distance from the shot point increases.

### 12.1.1 *The principal refractors*

P-wave velocity ranges for common rocks were indicated in Fig. 10.1, but in shallow refraction work a simple description of the ground in terms of dry overburden, wet overburden and weathered and fresh bedrock is often sufficient. It is very difficult to deal with situations involving more than three interfaces.

The P-wave velocity of dry overburden is often close to 350 m/sec, the velocity of sound in air, and is seldom more than 800 m/sec. There is usually a slow increase with depth, which is almost impossible to measure, followed by an abrupt increase to 1500–1800 m/sec at the water table.

Fresh bedrock usually has a P-wave velocity of more than 2500 m/sec, but is likely to be overlain by a transitional weathered layer where the velocity may be less than 2000 m/sec, usually increasing steadily with depth and the corresponding reduction in weathering.

Snell's Law (Section 10.1.5) implies that, if $v_2$ in Fig. 12.1 is greater than $v_1$, then at a critical angle of incidence given by:

$$\sin i = v_1/v_2 \qquad (12.1)$$

the refracted ray will travel parallel to the interface at velocity $v_2$.

Some energy returns to the ground surface after critical refraction, being represented by rays which leave the interface at the critical angle, defining a plane wavefront known as the *head wave*. The head wave travels through the upper layer at velocity $v_1$ but, because of its inclination, appears to move across the ground at the $v_2$ velocity with which the wavefront expands in the lower layer. It will therefore eventually overtake the direct wave, despite the longer travel path. The point at which the travel times for direct and refracted waves are equal is given by:

$$x_c = 2d \sqrt{\frac{v_2 + v_1}{v_2 - v_1}} \qquad (12.2)$$

$x_c$ is always more than double the interface depth and is large if the depth is large or the difference in velocities is small. It is known as the *crossover* or *critical distance*. The latter term is also sometimes used for the minimum distance at which refractions return to the surface, which is also the distance from the shot point at which energy is observed after reflection at the critical angle. This usage is not common amongst field crews since at this point, and for some distance beyond, the refraction arrives after the direct wave and is difficult or impossible to observe.

The critical time is obtained by dividing the crossover or critical distance by the direct-wave velocity.

If more than one interface is

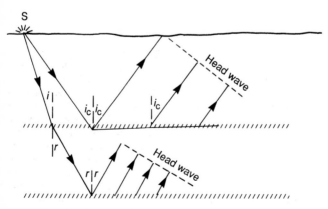

**Fig. 12.1** Critical refraction at two interfaces. $\sin i_c = v_1/v_2$, $\sin r = v_2/v_3$ and $\sin i = v_1/v_3$.

involved, the condition for critical refraction at the lowermost interface is given by the angle of the ray in the uppermost layer and the velocities in the uppermost and lowermost layers, i.e.

$$\sin i_n = v_1/v_n \qquad (12.3)$$

Although the other angles and velocities cancel out, the crossover equations become rather complicated and the intercept-time method of interpretation described in Section 12.2 is generally preferred.

### 12.1.3 Lengths of refraction spreads

A line of geophones laid out for a refraction survey is usually known as a *spread*, the term *array* being reserved for geophones feeding a single recording channel. Arrays are common in reflection work but almost unknown in refraction surveys where the sharpest possible arrivals are needed.

Sufficient information on the direct wave and reasonable coverage of the refractor is obtained if the length of the spread is about three times the crossover distance. A simple but often inaccurate rule of thumb states that the spread length should be eight times the expected depth.

### 12.1.4 Positioning shots

In most refraction surveys, some of the shots are fired very close to the ends of the spread. Interpretation is simplest if these shots are actually at the end-geophone positions so that times from other shots to the 'short-shot' points are recorded directly. The geophone should be moved from this position half-way towards the next in line when the short shot is fired from that end, to avoid damage and to obtain some extra information on the direct wave velocity.

'Long shots' may also be used. These should be placed sufficiently far from the spread to ensure that all first arrivals have come via the refractor, and short-shot data may therefore be needed before their positions can be decided. Offset distances need be measured only if continuous coverage is being obtained and the long-shot to one spread is to be in the same place as a short or centre shot to another. If explosives are being used, it may be worthwhile offsetting by a very considerable distance if this will allow firing in water (see Section 10.2.1).

### 12.1.5 Centre shots

The information provided by the conventional four-shot pattern may be supplemented by a centre shot. Centre shots are especially useful if there are considerable differences in interpretation at opposite ends of the spread, and especially if these seem to imply different numbers of refractors. They may make it possible to obtain a more reliable estimate of the velocity along an intermediate refractor and to monitor the thinning of an intermediate layer which is hidden, at one end of the spread, by refractions from greater depths. An additional reliable depth estimate is obtained which does not depend on assumptions about the ways in which the thicknesses of the various layers vary along the spread and there will be extra data on the direct wave velocity.

Centre shots are used less than they should be. The extra effort is generally trivial compared to the work involved in laying out the spread, and the additional and possibly vital information is cheaply obtained.

## 12.1.6 *Annotation of field records*

The primary records in most refraction surveys are produced in the field either photographically, or as in Fig. 12.2, by dot matrix printer. In a day's work which includes repeats, checks and tests as well as the shooting of a number of different spreads, several dozen records may be produced. They must be carefully annotated if confusion is to be avoided.

Annotations should obviously include the date and the name of the observer-in-charge, along with the survey location and spread number.

**Fig. 12.2** Nimbus 12-channel seismograph, showing dot-matrix 'hard-copy' emerging from the printer slot.

Orientation should be noted, and the position of Geophone 1 should be defined. Unless the geophone spacing is absolutely uniform, a sketch showing shot position and geophone locations should be added. If the interval between timing lines can be varied and/or variable time offsets can be applied, these settings must also be noted.

Other items may be considered optional. Amplifier gains and filter settings are not often recorded, but if filters are being used then such information may be useful. With enhancement instruments the number of shots or impacts combined in a single record can be important. And, of course, features such as the use of S-wave geophones at some points or peculiarities in the locations of some of the geophones should always be noted.

## 12.1.7 Picking refraction arrivals

First arrivals on refraction records may be difficult to identify at remote geophones where the signal-to-noise ratio is small (see Fig. 12.3). Some of the later peaks and troughs in the same wavetrain are likely to be stronger, and it is sometimes possible to work back from these to an estimate of the position of the first break. However, there are objections to this approach.

High frequencies are selectively absorbed in the ground and the distance between first break and any later peak therefore gradually increases with increasing distance from the source. Furthermore, the trace beyond the first break is affected by many other arrivals, as well as by later parts of the primary wavetrain, and these will modify peak and

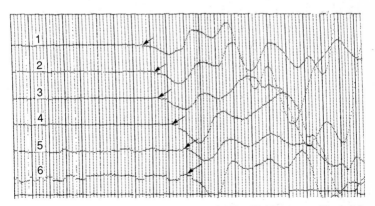

**Fig. 12.3** Six-channel refraction record, with 'picks' identified by arrows. These picks become more difficult to make with the decreasing signal-to-noise ratio on Channels 5 and 6. This record would be considered of high quality, and very much more difficult decisions usually have to be made.

trough locations. Using later features of traces to estimate first-arrival times should always be regarded as merely supplementary to direct picking.

### 12.1.8 Time–distance plots

The raw data from refraction surveys consist of sets of times (usually first-arrival times) measured at geophones at various distances from the source positions. These are plotted against vertical time axes and horizontal distance axes. The gradient of any line will be equal to the reciprocal of a velocity so that steep slopes correspond to slow velocities. All data for a given spread are plotted against the same axes and without regard for source position (see Fig. 12.8 on p. 173).

The plot has a working area from the first to the last geophone only and need not be extended out to show the long-shot positions. There can be as many as five sets of arrivals to be plotted, as well as a set of time differences. Different colours or symbols should be used to distinguish different data sets.

If the arrival times lie on a number of clearly defined straight-line segments, best-fit lines may be drawn. However, these are not actually necessary when long shots are being used and will be difficult to draw if there is considerable relief on the refracting surfaces, because arrival times will be irregular. It is often best if lines are drawn only through the direct-wave arrivals (which should plot on straight lines), leaving refrac-ted arrivals either unjoined or linked only by dotted lines between adjacent points.

## 12.2 Field interpretation

Interpretation is an essential part of refraction fieldwork because the success of a survey depends on parameters such as line orientation, geophone lay-outs, shot positions and spread lengths, which can be varied almost at will. Only if analysis keeps pace with the survey work will the right choices be made.

Depths estimated from refraction surveys are related to geophone and shot-point elevations, which must therefore be measured by conventional land-survey techniques to obtain a true picture of the sub-surface refractor.

### 12.2.1 Intercept times

The intercept time, $T_i$, is defined as the time at which the back-extrapolated refracted arrival line cuts the time axis. For a single refractor, $T_i$ is related to the velocities and the refractor depth by the equation:

$$T_i = \frac{2d\sqrt{(v_2^2 - v_1^2)}}{v_1 v_2} = \frac{2d}{v_{1,2}} \quad (12.4)$$

The quantity $v_{1,2}$ is defined by this equation. It has the units of a velocity and is approximately equal to $v_1$ if $v_2$ is very much larger than $v_1$. The critical angle is then almost 90° and the delay suffered by the refracted ray in travelling between the surface and the refractor is close to double the vertical travel time. $v_{1,2}$ is large if the difference between $v_1$ and $v_2$ is small.

Intercept times are conventionally obtained by drawing best-fit lines through the refracted arrivals. However, even if the fit is very good, there is no guarantee that the depth of the refractor does not change in the region near the shot point where refractions are not observed. If a long shot is used, there should be a constant difference between long-shot and short-shot arrival times at points towards the far end of the spread (Fig. 12.4). An intercept time can then be obtained by subtracting this difference from the long-shot arrival time at the short-shot location. This can be done exactly if there is geophone in this position when the long shot is fired (see Fig. 12.8). Otherwise, use of the nearest long-shot arrival at least reduces the distance over which extrapolation must be made.

### 12.2.2 Multiple layers

Equation (12.4) can be extended to cases involving a number of critically refracting layers. If the intercept time associated with the $n$th refractor is $T_n$, then:

$$T_n = \frac{2d_1}{v_{1,n+1}} + \frac{2d_2}{v_{2,n+1}} \ldots + \frac{2d_n}{v_{n,n+1}} \quad (12.5)$$

where $d_n$ is thickness of the $n$th layer which overlies the $n$th refracting interface, at which velocity increases from $v_n$ to $v_{n+1}$. The presence of intermediate layers may be recognized by comparing long- and short-shot data (see Fig. 12.4), but at least two points are needed to define a velocity and three for any confidence to be placed in the estimate. At the best, therefore, only four layers can be easily investigated with a 12-channel system.

Complicated field procedures can be devised to overcome this limitation; geophones may, for example, be moved one half-interval after a shot has been fired and the same shot-point can then be re-used. Progress is extremely slow, and the problems presented by refractor topography, hidden layers and blind zones

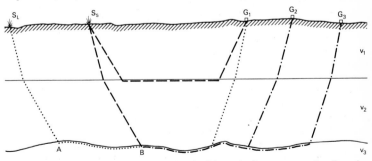

**Fig. 12.4** Long-shot and short-shot travel paths. The paths from source points $S_L$ and $S_s$ are identical from the point B onwards.

(Section 12.3) still exist. In most circumstances, multiple shots into modified spreads represent attempts to extract more from the method than is really obtainable.

## 12.2.3 Effect of dip

As far as seismic methods are concerned, the horizontal is an arbitrary reference; the 'depths' determined are the perpendicular distances from shot points or geophones to interfaces. With this proviso, 'horizontal' formulae can be applied without modification where the ground surface and the refractor dip at the same angle. More usually the slopes will be different, and the formulae are then most commonly quoted in terms of a horizontal ground and a dipping refractor. Equation (12.4) gives the perpendicular distance from the shot point to the refractor

provided that the true value of $v_2$ is used.

$v_2$ may not be readily available. A wave which travels down-dip not only has to travel further at the $v_2$ velocity to reach more distant geophones, but also further at velocity $v_1$ through the upper layer, and so arrives 'late' (Fig. 12.5). The reverse is true for an up-dip spread, where arrivals at further geophones may actually precede those at nearer ones. The slope of the line through the refracted arrivals on a time–distance plot depends on the dip $\alpha$ according to:

$$v_{\mathrm{app}} = \frac{v}{1 + \sin \alpha} \qquad (12.6)$$

If the geophones are offset from the source in the downdip direction, $\alpha$ is positive. If shots are fired from both ends of the spread, different apparent velocities will be measured because the sign of $\alpha$ will differ. The true

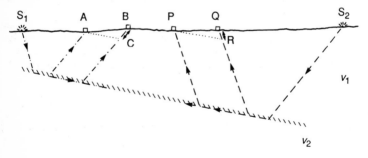

**Fig. 12.5** Refraction at a dipping interface. The refracted energy from $S_1$ arrives later at B than at A not only because of the greater distance travelled along the interface but also because of the extra distance CB travelled in the low velocity material. Energy from $S_2$ arrives earlier at P than would be expected, given its distance from Q, by the time taken to travel RQ at the $v_1$ velocity. (Note: The dotted lines AC and PR are parallel to the refractor).

168

velocity, if the dip angle is less than about 10°, is given by the equation:

$$\frac{2}{v} = \frac{1}{v_{up}} + \frac{1}{v_{down}} \qquad (12.7)$$

### 12.2.4 Refractor relief and true velocities

Most refractors except the water table are irregular.

If there were only a single local depression in an otherwise flat refractor, the refracted arrivals from the two ends would plot on straight lines of equal slope, except at the geophone immediately over the depression, and their differences would plot on a line with double this slope. Above the depression the waves from both ends would arrive late (Fig. 12.6), and moreover the delays for small dips would be very similar. The difference between the arrival times would thus be almost the same as if no depression existed, plotting on the straight line generated by the horizontal parts of the interface.

This argument can be extended to a refractor which forms a series of depressions and highs, provided the dip angles are low. The difference points plot along a straight line with slope corresponding to half the refractor velocity. The approach generally works better than this very qualitative 'proof' might suggest, with changes in slopes of difference lines corresponding to real changes in refractor velocity. Zones of bedrock weakness can be located using this technique.

The importance of long shots is obvious, since the part of the spread over which the short shot first arrivals from both ends have come via the refractor is likely to be rather short and may not even exist. Centre shots sometimes allow the same technique to be applied to determining the velocity along an intermediate refractor.

The slope of the difference line over a refractor with constant dip will equal the sum of the slopes of the individual lines. This is equation (12.7) in graphical form.

Differences are easily obtained directly from the plot using dividers or a straight-edged piece of paper, and are plotted on the same axes. An

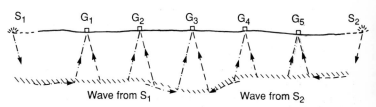

**Fig. 12.6** Effect of a bedrock depression on travel times. The arrivals at $G_3$ of energy from $S_1$ and $S_2$ are delayed by approximately the same amount.

arbitrary time zero is used, placed where there will be the least overlap with other data (see Fig. 12.8).

### 12.2.5 Reciprocal time interpretation

The *reciprocal time*, $t_R$, is the time taken for seismic energy to travel from one long-shot position to the other. The difference between $t_R$ and the times $t_A$ and $t_B$ taken by the energy from the long shots to arrive at a given geophone is directly related to the depth of the refractor beneath the interface (Fig. 12.7).

$$t_A + t_B - t_R = 2d/F \qquad (12.8)$$

If there is only a single interface, so that the total depth $D$ is equal to the thickness $d$ of the upper layer, the depth conversion factor $F$ is equal to $v_{1,2}$. If more than two layers are involved, $F$ is a composite of all the velocities involved, weighted according to the layer thicknesses. It can be calculated by applying equation (12.8) at the short-shots, where the refractor depths are known from the intercept times. The way in which it varies between shot points may be very complex, but for field interpretation it is enough to interpolate linearly (Example 12.1). $t_R$ may be measured directly, but this is seldom convenient. It can also be calculated, provided that geophones are located at the short-shot points when the long shots are being fired, since at these points the right-hand side of equation (12.8) is equal to the intercept time. The calculated reciprocal times should be the same in both directions. If this is not so, the raw data and the calculations must be thoroughly checked to find the reason for the discrepancy.

Short-shot reciprocal times are measured directly if the shots are fired from the end-geophone position, and the fact that they should be equal may help in picking arrivals. However, they have little interpretational significance.

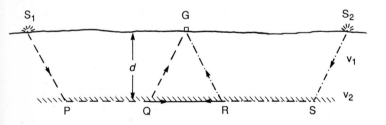

**Fig. 12.7** Reciprocal time interpretation. The sum of the travel-times to G from $S_1$ and $S_2$ differs from the reciprocal time $t_R$ (the time taken to travel from $S_1$ to $S_2$) only by the difference between the times to travel the paths QR (at velocity $v_2$) and QGR (at velocity $v_1$).

## 12.3 Limitations of the refraction method

First-arrival refraction work uses only a small proportion of the information contained in the seismic traces, and it is not surprising that interpretations are subject to some severe limitations. These are especially important in engineering work; in low-velocity-layer studies only a time delay estimate is sought and short shots alone are often sufficient.

### 12.3.1 Direct waves

The 'ground roll' consists of a complex of P- and S- body waves and Love and Rayleigh surface waves which travel with different but generally slow velocities. There is often some doubt as to which component actually produces the first break, since conventional geophones respond only poorly to the horizontal ground motions produced by direct P-waves. Close to the source enough energy is associated with the P-waves for the response to be measurable, but at greater distances the first breaks may record the arrival of S-waves, surface waves or even the air wave. Conventionally-connected geophones 'break' downwards for refractions but may break either way for direct waves.

The complex character of the direct wave may be among the reasons for the failure of many direct-wave arrival lines to pass through the origin. Delays in the timing circuits also play their part, but these can be determined by direct experiment, with a detonator or a light hammer blow immediately next to a geophone. Gains at geophones close to a shot point are often set so low that the true first-arrivals are overlooked (Fig. 12.9). If this is the case, then the reliable velocity estimates will be those which do not treat the origin as a point on the plot.

### 12.3.2 Vertical velocities

However much care is taken to obtain valid direct-wave or refracted-wave velocities, the refraction method is fundamentally flawed in that measured horizontal velocities are used in depth equations which require vertical velocities. If there is significant anisotropy, errors will be introduced. This is more a problem for the interpreter than the field observer, but the latter should at least be aware of the importance of using any boreholes or recent excavations to calibrate interpretation or even to measure vertical velocities directly.

### 12.3.3 Hidden layers

A refractor which does not give rise to any first-arrivals is said to be *hidden*. A layer is likely to be hidden if it is thin in comparison with the layer above and has a much lower seismic velocity than the layer below. Weathered layers immediately over basement often fulfil these criteria.

# Example 12.1

---

Field interpretation of 4-shot refraction spread with long shot (LS) and short shot (SS) arrivals from West (W) and East (E) ends plotted on same set of axes (Figure 12.8).

*Stage 1 (Base refractor intercept times)*

Measure WLS − WSS time differences. These are close to 41 msec from G6 to G12, indicating region where WSS arrivals have come from base refractor.
   WLS time at W end = 101 msec. Intercept time = 101 − 41 = 60 msec.
   Similarly, ELS − ESS time differences are close to 59 msec from G1 to G4.
   ELS time at E end = 208 msec. Intercept time = 208 − 59 = 149 msec.
(Note extrapolated intercept time of about 170 msec is very different)

*Stage 2 (velocities)*

(a) Straight line from W origin through nearby WSS arrivals extends 60 m to G4. Velocity = $60/0.079 = 759$ m·sec$^{-1}$
   Straight line from E origin through nearby ESS arrivals extends 100 m to G7. Velocity = $100/0.134 = 746$ m·sec$^{-1}$
   Average $v_1$ value = 750 m·sec$^{-1}$

(b) Arrivals at G5 (WSS) and at G5 and G6 (ESS) belong neither to 'base refractor' sets (see Stage 1) nor to '$v_1$' sets, suggesting presence of an intermediate, '$v_2$', refractor. $v_2$ slope lines are poorly controlled but should pass above all direct wave ($v_1$) and base refractor first arrivals. For the most likely positions, as shown:
   WSS: $v_2 = 1470$ m·sec$^{-1}$ Intercept time = 29 msec
   ESS: $v_2 = 1560$ m·sec$^{-1}$ Intercept time = 77 msec
   (Interface is probably water table, velocity about 1500 m·sec$^{-1}$)

(c) Measure and plot WLS − ELS time differences at each geophone, using any convenient line as time zero.
   $v_3 = 2/(\text{slope of difference line}) = 2 \times 220/0.182 = 2420$ m·sec$^{-1}$

(d) $v_{1,2} = v_1 \times v_1/\sqrt{(v_2^2 - v_1^2)} = 750 \times 1500/\sqrt{(1500^2 - 750^2)} = 870$ m·sec$^{-1}$
$v_{1,3} = 750 \times 2420/\sqrt{(2420^2 - 750^2)} = 790$ m·sec$^{-1}$;
$v_{2,3} = 1500 \times 2420/\sqrt{(2420^2 - 1500^2)} = 1910$ m·sec$^{-1}$

*Stage 3 (Depths at shot points)*

Depth to intermediate refractor = $d_1 = \frac{1}{2} t_i\, v_{1,2}$
W end: $d_1 = \frac{1}{2} \times 0.029 \times 870 = 12.6$ m; E end: $d_1 = \frac{1}{2} \times 0.077 \times 870 = 33.5$ m
   Thickness of second layer = $d_2 = \frac{1}{2} \times [t_i - 2d_1/v_{1,3}] \times v_{2,3}$
W end: $d_2 = \frac{1}{2} \times [0.060 - 25.2/790] \times 1910 = 26.8$ m
$D = 26.8 + 12.6 = 39.4$ m

E end: $d_2 = \frac{1}{2} \times [0.149 - 67.0/790] \times 1910 = 61.3$ m
$D = 33.5 + 61.3 = 94.8$ m

*Stage 4 (Reciprocal time interpretation – example using G8)*
Reciprocal time $= t_R = $ (WLS time + ELS time) − (intercept time)
W end: $t_R = 101 + 254 - 60 = 295$ msec; E end: $t_R = 233 + 208 - 149 = $
292 msec

Average of reciprocal time estimates $= 293$ msec $= 0.293$ secs
Depth conversion factor $F$ at a short shot $= 2 \times D \div$ (intercept time)
West: $F = 2 \times 39.4/0.060 = 1310$ m·sec[1];
East: $F = 2 \times 94.8/0.149 = 1270$ m·sec$^{-1}$
$F$ at G8 (by interpolation) $= 1280$ m·sec$^{-1}$
WLS time at G8 $= 0.174$ sec; ELS time $= 0.213$ sec; Sum $= 0.387$ sec
$D = (0.387 - 0.293) \times 1280/2 = 60.2$ m

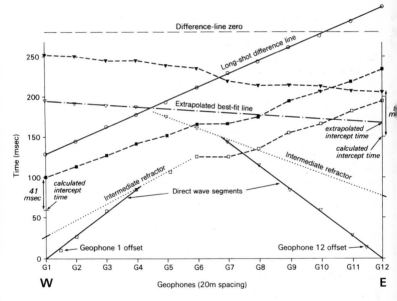

**Fig. 12.8** Time–distance plot for a four-shot refraction spread. The short shot arrivals from the West end (WSS) are indicated by open squares and the long shot arrivals (WLS) by solid squares. ESS and ELS arrivals are indicated by open and solid triangles respectively. The difference line (open circles) is referred to a zero line arbitrarily positioned at 280 msecs. Note locations of intercept times for the 'most likely' extrapolated intermediate refractor lines.

173

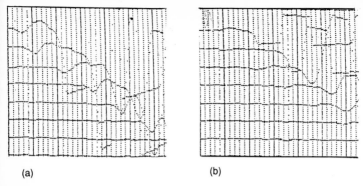

**Fig. 12.9** Result of playing-back the same refraction data with different amplification. The first arrivals clearly visible in (*a*) would probably be overlooked or dismissed as noise on (*b*). A direct-wave velocity based on the later arrivals on (*b*) would be roughly correct as long as the best-fit line was not forced through the origin. The crossover distance would also be wrong, but the intercept time would not be affected if the amplification were sufficient for the refracted first-arrivals to be seen.

The presence of a hidden layer can sometimes be recognized by studying second arrivals with intermediate velocity, but since refracted waves are strongly attenuated in thin layers, this is only occasionally possible.

A layer may be hidden, even if the head wave from it does arrive first over some part of the ground surface, if there are no appropriately located geophones. Smaller geophone spacing in the critical region can sometimes be useful but the necessity will only be recognised if preliminary interpretations are being made on a daily basis.

*12.3.4 Blind zones*

If velocity decreases at an interface, critical refraction cannot occur and no refracted energy will return to the surface. Little can be done about such interfaces, said to be *blind*, unless vertical velocities can be measured directly.

Thin high-velocity layers such as perched water tables and buried terraces often overlie blind zones. The refracted waves within them lose energy rapidly with increasing distance from the source and ultimately become undetectable. Much later events may then be picked as first arrivals, producing an abrupt jump in the time–distance plot. Disappearance of the refracted arrivals may also be due to the actual pinching out of the layer responsible.

*12.3.5 Limitations of drilling*

Despite the problems with refraction surveys, it is not always the seismic

174

interpretation that is wrong when it differs from drillhole data. Only a very small sub-surface volume is sampled by the drill, and many tests of drift thickness have been terminated in isolated boulders some distance above the true top-of-bedrock. All possible reasons for differences between drillhole and seismic data must be considered.

# Appendix

## *Terrain corrections for Hammer zones B to M*

| Zone: | B | C | D | E | F | G |
|---|---|---|---|---|---|---|
| No. of compartments: | 4 | 6 | 6 | 8 | 8 | 12 |
| Correction (g.u.): | | | Heights (metres) | | | |
| 0.01 | 0.5 | 1.9 | 3.3 | 7.6 | 11.5 | 24.9 |
| 0.02 | 0.7 | 2.6 | 4.7 | 10.7 | 16.3 | 35.1 |
| 0.03 | 0.8 | 3.2 | 5.8 | 13.1 | 19.9 | 13.1 |
| 0.04 | 1.0 | 3.8 | 6.7 | 15.2 | 23.0 | 49.8 |
| 0.05 | 1.1 | 4.2 | 7.5 | 17.0 | 25.7 | 55.6 |
| 0.06 | 1.2 | 4.6 | 8.2 | 18.6 | 28.2 | 60.9 |
| 0.07 | 1.3 | 5.0 | 8.9 | 20.1 | 30.4 | 65.8 |
| 0.08 | 1.4 | 5.4 | 9.5 | 21.5 | 32.6 | 70.4 |
| 0.09 | 1.5 | 5.7 | 10.1 | 22.9 | 34.5 | 74.7 |
| 0.10 | 1.6 | 6.0 | 10.6 | 24.1 | 36.4 | 78.7 |
| 0.20 | 2.4 | 8.7 | 15.1 | 34.2 | 51.6 | 111.6 |
| 0.30 | 3.2 | 10.9 | 18.6 | 42.1 | 63.3 | 136.9 |
| 0.40 | 3.9 | 12.9 | 21.7 | 48.8 | 73.2 | 158.3 |
| 0.50 | 4.6 | 14.7 | 24.4 | 54.8 | 82.0 | 177.4 |
| 0.60 | 5.3 | 16.5 | 26.9 | 60.2 | 90.0 | 194.7 |
| 0.70 | 6.1 | 18.2 | 29.3 | 65.3 | 97.3 | 210.7 |
| 0.80 | 6.9 | 19.9 | 31.5 | 70.1 | 104.2 | 225.6 |
| 0.90 | 7.8 | 21.6 | 33.7 | 74.7 | 110.8 | 239.8 |
| 1.00 | 8.7 | 23.4 | 35.7 | 79.1 | 117.0 | 253.2 |

|      | H 12 | I 12 | J 16 | K 16 | L 16 | M 16 |
|------|------|------|------|------|------|------|
| 0.01 | 32 | 42 | 72 | 88 | 101 | 125 |
| 0.02 | 46 | 60 | 101 | 124 | 148 | 182 |
| 0.03 | 56 | 74 | 125 | 153 | 186 | 225 |
| 0.04 | 65 | 85 | 144 | 176 | 213 | 262 |
| 0.05 | 73 | 95 | 161 | 197 | 239 | 291 |
| 0.06 | 80 | 104 | 176 | 216 | 261 | 319 |
| 0.07 | 86 | 112 | 191 | 233 | 282 | 346 |
| 0.08 | 92 | 120 | 204 | 249 | 303 | 370 |
| 0.09 | 96 | 127 | 216 | 264 | 322 | 391 |
| 0.10 | 103 | 134 | 228 | 278 | 338 | 413 |
| 0.20 | 146 | 190 | 322 | 394 | 479 | 586 |
| 0.30 | 179 | 233 | 396 | 483 | 587 | 717 |
| 0.40 | 206 | 269 | 457 | 557 | 679 | 828 |
| 0.50 | 231 | 301 | 511 | 624 | 759 | 926 |
| 0.60 | 253 | 330 | 561 | 683 | 832 | 1015 |
| 0.70 | 274 | 357 | 606 | 738 | 899 | 1097 |
| 0.80 | 293 | 382 | 648 | 790 | 962 | 1173 |
| 0.90 | 311 | 405 | 688 | 838 | 1020 | 1244 |
| 1.00 | 328 | 427 | 726 | 884 | 1076 | 1312 |

**Note** These tables list the exact height differences which, assuming a density of $2000 \, kg \cdot m^{-3}$, will produce the tabulated terrain effects. Thus, a height difference of 32 m between gravity station and average topographic level in one compartment of Zone E would be associated with a terrain effect of 0.20, or possibly 0.19, g.u. Almost all commercial gravity meters have sensitivities of 0.1 g.u. but an additional decimal place is necessary if large 'rounding off' errors are to be avoided in summing the contributions from all the compartments. The inner radius of Zone B is 2 m. Zone outer radii are: B: 16.6 m, C: 53.3 m, D: 170 m, E: 390 m, F: 895 m, G: 1530 m, H: 2.61 km, I: 4.47 km, J: 6.65 km, K: 9.9 km, L: 14.7 km, M: 21.9 km.

# References and further reading

BARKER, R. D., (1981) 'The offset system of electrical resistivity sounding and its use with a multicore cable.' *Geophysical Prospecting*, 29, 128–143.

CORWIN, R. F. & HOOVER, D. B., (1979) 'The self-potential method in geothermal exploration.' *Geophysics*, 44, 226–245.

COUTTS, D. A., WELLMAN, P. & BARLOW, B. C., (1984) 'Calibration of gravity meters with a quartz mechanism.' *BMR Jnl. Aust. Geol. Geophys.*, 5, 1–7.

FRASER, D. C., (1969) 'Contouring of VLF data.' *Geophysics*, 34, 958–967.

GRIFFITHS, D. H. & KING, R. F., (1981) *Applied Geophysics for Geologists and Engineers*. Oxford, Pergamon Press, 230 pp.

HAMMER, S., (1939) 'Terrain corrections for gravimeter stations.' *Geophysics*, 4, 184–194.

HAWKINS, L. V., (1961) 'The reciprocal method of routine shallow seismic refraction investigations.' *Geophysics*, 26, 806–819.

IMPERIAL CHEMICAL INDUSTRIES, (1962) *Blasting Practice*. Ayrshire, ICI Nobel Division, 270 pp.

KEAREY, P. & BROOKS, M., (1984) *An Introduction to Geophysical Exploration*. Oxford, Blackwell Scientific Publications, 296 pp.

LONGMAN, I. M., (1959) 'Formulas for computing the tidal accelerations due to the moon and sun.' *J. Geophys. Res.*, 64, 2351–2355.

LILLEY, F. E. M. & BARTON, C. E. (eds), (1986) *Geomagnetism in an Australian Setting*. *Exploration Geophysics* (special edition). Sydney, Australian Society of Exploration Geophysicists, 58 pp.

NETTLETON, L. L., (1976) *Gravity and Magnetics in Oil Prospecting*. New York, McGraw-Hill, 464 pp.

PARASNIS, D. S., (1975) *Mining Geophysics*. Amsterdam, Elsevier, 395 pp.

QUICK, D. H., (1974) 'The interpretation of gradient array chargeability anomalies.' *Geophysical Prospecting*, 22, 736–746.

RIDDIHOUGH, P., (1971) 'Diurnal corrections to magnetic surveys—an assessment of errors.' *Geophysical Prospecting*, 19, 551–567.

STALTARI, G., (1986) 'The Que River TEM case-study.' *Exploration Geophysics*, 17, 125–128.

SUMNER, J. S., (1976) *Principles of Induced Polarisation in Geophysical Exploration*. Amsterdam, Elsevier, 277 pp.

TELFORD, W. M., GELDART, L. P., SHERIFF, R. E. & KEYS, D. A., (1976) *Applied Geophysics*. Cambridge, Cambridge University Press, 841 pp.

WHITELY, R. J. (ed.), (1981) *Geophysical Case Study of the Woodlawn Orebody, New South Wales, Australia*. Oxford, Pergamon Press, 588 pp.

# Index

182